How to Become a Successful Driving Instructor

Orders: You can order through Amazon.co.uk under ISBN 9781910662656

ISBN: 9781910662656

First published November 2018

 All costs shown were correct as of November 2018.

CONTENTS

PREFACE

Hi and a warm welcome…

You're probably wondering who I am and why write a book. So let me give you a quick potted history:

I'm a fully qualified approved driving instructor (ADI), which means I spend much of my time sitting in the passenger seat of my car teaching mainly 17 – 20 year olds to drive. I also teach people from all walks of life and of all ages to become driving instructors. This gives me a real buzz, as it's life-changing for my pupils and every pupil is unique.

Back in 2014 I was where you probably are now. Toying with the idea of becoming a driving instructor. So I started by simply doing a Google search "become a driving instructor" and bingo, I was on my way. What I discovered in the months and years that followed is that the harsh realities of life as a driving instructor are very different to what I had read about in the glossy brochures that had landed on my doormat.

Since then I've sat in the back of around 200 driving tests and spoken at length with over 20 other driving instructors about their own experiences. I've also trained others from all walks of life, young and old to become driving instructors.

So that's the main reason for writing the book: to give you an honest and true account of my own journey and experiences, which hopefully you will benefit from. I'll tell you what is great about being a driving instructor and what traps to look out for, based on my own experiences. In short, I'll explain what I would do now if I was starting out again with the benefit of hindsight.

What I have written in this book is very current, as my own training was conducted only four years ago. It's based on the new ADI part 3 test, which was introduced at the start of 2018, so it's completely relevant to what you will encounter should you embark on the journey.

I changed my instructor three times when qualifying, as I just wasn't at all happy with the quality of the training I was being given. My final trainer, who steered me through my ADI part 3 test (after 2 failed attempts with other instructors) was simply brilliant. He was formerly the head of examiner recruitment and training at the DVSA training headquarters and he taught me exactly the same as he had taught the examiners! His training was in a completely different league and I passed my part 3 test on my third and final attempt with a near perfect score. The quality of the training varies enormously and is absolutely key to your success. I will expand on this at length in the chapters that follow.

The word 'successful' in the title of the book takes on two meanings. Firstly how to successfully qualify as an ADI and secondly how to successfully run your own driving instructor business. And remember that it will always be your own business, no matter which route you go down.

By the end of the book, you will hopefully have a greater understanding of the realities of both the training that's involved and also what life is really like at the end of the tunnel.

If you're seriously thinking about coming into the industry, this book will definitely help you make an informed decision about which path to go down.

Finally, I welcome any feedback or questions. Please feel free to email me at bc@briancroucher.co.uk and I'll do my best to help.

Enjoy…

Congratulations Chelsea on being my 100th pupil to pass their test.

Very well done to the other 99 as well!

A Few Facts about the Industry.....

There are 220 driving test centres in the UK (as of Nov 2018).

1,910,169 driving tests were carried out in the year ending 31st March 2018.

This number is 1.3% lower than in the previous year.

465,650 practical driving tests were carried out in the 3 months from 1st January 2018 to 31st March 2018. This compares to 395,640 tests carried out during the same period in 2013. This shows an increase of 17.7% over 5 years.

On 31st March 2018 there were 39,348 approved driving instructors (ADI's) in the UK registered with the DVSA. This is 0.4% lower than in March 2017 and 12% lower than in March 2013.

So over the last 5 years...

- The number of practical driving tests conducted has increased by 17.7%, but
- The number of ADI's has fallen by 12%.

Over 50% of all people who commence their training to become driving instructors either drop out or fail to qualify.

The pass rate in 2017 for ADI Part 3 tests was a mere 27%.

CHAPTER 1 – Nerves of Steel?

“Hi Brian, nice to meet you. And what do you do for a living?”

“I’m a driving instructor”.

“Oh really, you must have nerves of steel”.

I’ve lost count of the number of people I’ve had such a conversation with. But do you really need nerves of steel?

I certainly needed nerves of steel when I taught my own children to drive. But that was in my own family car, which didn’t have dual controls. The car I teach in now (a Peugeot 208 HDI) is fitted with industry-standard “He-Man” dual controls, which give me separate brake and clutch pedals. This means that I still have some control of the car even though I’m sitting in the passenger’s seat. At first it’s a bit daunting, but you soon get used to it. So no, I don’t think you need nerves of steel to become a driving instructor.

I’ve taught people from all walks of life to become driving instructors, including

- Teacher
- Retail store manager
- Taxi driver
- Bus driver
- Account manager
- Delivery driver
- Tree surgeon
- Plasterer
- University graduate.

Age range from 22 to 63. Male and female. In my experience this makes very little difference.

Many people want to make driving instruction their full-time career. But the hours you work are completely up to you. Some instructors wrap their hours around an existing job by working evenings and weekends. Others use driving instruction as a way of topping up their pension. Maybe they’ve taken early retirement and only want part-time work. Say 3 days per week, or 5 mornings per week. Perfect.

I know one driving instructor who only works during British Summer Time from March until the end of October. Then he goes travelling in the southern hemisphere and misses the UK winter. Brilliant! That's what being your own boss is all about: you have complete control of your time and the freedom to work when it suits you.

By far the biggest factor in my opinion is your **Why.** Why do you want to become a driving instructor? What is your compelling reason?

So what sort of person makes a great driving instructor?

The most important qualities in my opinion are:

Why. What is your 'why'? You why is a huge factor. If you have recently been made redundant for example then you have a huge why, as you need to replace your income. Similarly, if you hate your current job then the benefits of starting a new career are huge, so your 'why' will also be huge. You'll do whatever it takes to be successful, because the pain of not being successful is so great.

On the other hand, if you're not 100% sure that this is right for you, you may decide to fit in your training around your current job. Therefore your 'why' is probably not as big as if you were totally committed. As a result you're less likely to be successful in your training.

Take a moment now to write down at least five reasons why you want to become a driving instructor and what these will mean to you.

For example:

Be my own boss. The benefit is that I choose when I work, so I can spend time my children in the morning before they go to school.

Think of at least five different reasons why you want to become a driving instructor. The more, the better.

One of the compelling reasons for me **was** that two of my children live with their own families in Australia. I am determined to be a face-to-face grandad and not a FaceTime grandad. I am prepared to work long hours through the summer months so that I can have long holidays with my family in Australia in January and February each year (i.e. their summer).

Patient. Tell a young lady pupil on average 1.5 times and she'll remember it. On the other hand, the average number of times you need to tell a guy is about 8. Then you get a grunt and he'll say "why didn't you tell me that before?" I just keep telling myself that it's not them. I just haven't found the best way to connect with them. In a handful of cases where pupils just couldn't grasp the concept of gears, clutch and biting point I have transferred them over to learn in an automatic car with another instructor (we have a reciprocal arrangement and it works well). You have to do whatever is best for the pupil.

Organised. Diary management, pupils' lesson record keeping, phone numbers, pupil's home address, work address, lesson payment details, test date, car expenses. And so on. How you do it doesn't really matter. Many instructors (including myself) prefer paper and pen. Others keep an ipad to hand.

Good Communicator. Good communication skills are essential. Some pupils are verbal (ie you can tell them something and they'll remember it), others like to see pictures of junction diagrams, others are more practical (i.e. they need to actually experience it for themselves). I keep in touch with my pupils by text. I text them the evening before their lesson just to remind them. This works well.

Calmness. Pupils sometimes stall at busy junctions. When they do, it's important to stay calm yourself and exude calmness. Just talk them through what they need to do in a calm manner and use the dual controls if needs be. Then pull them over on the left at a safe place and explain what why they stalled the car and what they need to do differently in the future in order to prevent it from happening again.

Professional. You absolutely must have a professional approach towards your pupils. Imagine there's a glass divide running down the centre of your car. Know the boundaries. Stay on your side. Also, take care with the language that you use when texting pupils. I always sign off with imojis of a red car and a green tick so that the pupils know that the message relates to their driving lesson. One quick tip: never text "I'll pick you up at 9am tomorrow morning".

Instead, word it "I'll collect you at 9am tomorrow morning". I've had one instance of a jealous boyfriend thinking I was picking up his girlfriend! Lesson learned.

Firm but fair. I give all my pupils a very brief agreement, which sets out my cancellation policy. If a pupil cancels their lesson and gives me less than 24 hours notice then I have made it clear in black and white that I will try and fill their gap with another pupil, but if I can't then the lesson is deducted from the hours they have pre-paid for. It's amazing how many pupils make a dramatic recovery when they realise that they're going to lose their money if they cancel!

Positive. A positive can-do attitude is essential for confidence building. Constructive feedback is essential. Destructive criticism can destroy a pupil's confidence.

So I find at least one thing that the pupil did well before hitting them with the bad bit. For example, if the pupil stalled whilst moving off, I would say something like:

"really good observations there, so well done. The bit that we need to work now is finding the biting point".

Another thing I do which always works well on the pupil's first lesson is to ask them how they're feeling. Whatever they say next you then reply "that's great".

For example…

> "How are you feeling?"
>
> "A bit nervous".
>
> "That's great. It shows this means a lot to you. Most people I have taught felt the same way as you do now when they first started. You'll be fine".

Enthusiastic. Enthusiasm and high energy levels are contagious. Keep your energy levels high with regular 'pit stops', drink plenty of water and snack frequently. Talk with your pupils about whatever they enjoy talking about. This helps maintain their enjoyment and enthusiasm for driving. It also helps their minds relax. Also try having the radio or music on in the background occasionally and see if that helps them.

CHAPTER 2 –
The Two Different Ways to Qualify as a Driving Instructor: A Step-by-Step Guide

In this section I will explain step-by-step the two different ways that you can qualify as a driving instructor. I've broken it down into 2 phases…

Phase 1 is common to both ways of qualifying, i.e. you have to do this bit.

There are two different ways you can complete phase 2. I've called these Phase 2a and Phase 2b.

So to qualify as a driving instructor you have to complete and pass phase 1 and then either phase 2a or phase 2b. Confused? Then read on and all will be revealed.

Phase 1 (The same for both options)

Step 1 : Check to see if you're eligible

Go to Google and search for "become a driving instructor".

Then Click on the following .gov.uk website.

Check if you can become a driving instructor - GOV.UK

https://www.gov.uk/become-a-driving-instructor

Click on "check you can become a driving instructor"

Are you aged 21 or over?

Have you had a full car driving licence for at least 3 years?

What type of car is your driving licence for? (Manual or automatic)

The website will then tell you if you're eligible to apply. Your application may be refused if you have

- been banned from driving
- 6 or more penalty points on your licence
- been convicted of any non-motoring offences
- been banned from working with children

Step 2 : Get a DBS check (costs £6)

Follow the link on the .gov.uk website.

You will need to enter the ORG PIN (use '105205')

the secret word (use 'axis')

Follow the instructions on the website and order your DBS check.

This takes up to 2 weeks. You will then need to have your proof of ID and residence verified at a post office, for which there is a small charge.

You can then apply to the DVSA for your PRN (personal reference number).

Once you have your PRN, you can proceed to step 3.

Step 3 : Prepare for your ADI part 1 test (theory)

Your training organisation should provide you with books and online resources to prepare you for your ADI Part 1 test (theory).

There is a whole raft of online resources and apps available. The one I particularly like is (as it's set up for ADI part 1 test) is:

www.drivingtheory4all.co.uk

Step 4 : Book your ADI Part 1 theory test (costs £81)

Visit the following website:

Approved driving instructor (ADI) part 1 test - GOV.UK

https://www.gov.uk/adi-part-1-test

You can **book your** approved **driving** instructor (**ADI**) **part 1 test** when **your** application to ... Rebook **your ADI part 1 test** if you failed **your test** and want to resit it.

Book your instructor theory test · Revision and practice · What to take to your test

The theory test is split into two parts, namely:

- 100 multiple choice questions, and
- 14 hazard perception video clips

You have to pass both parts to be successful.

You have unlimited attempts to pass. The vast majority of my pupils pass first time. Occasionally they need two attempts, but unlike the ADI part 2 and part 3 tests, you can have an unlimited number of attempts, so there's no pressure.

Once you've passed your ADI part 1, the clock starts ticking. You've got 2 years to complete your training and pass your ADI part 2 and part 3 exams.

Step 5 : Book and pass your ADI Part 2 practical driving test (costs £111)

I suggest you book your ADI part 2 practical driving test as soon as you've passed your ADI part 1 (unless your trainer tells you otherwise), as waiting times for part 2 tests can be up to 2 months.

Most training organisations offer between 8 and 10 hours in-car tuition to prepare pupils for their part 2 test. This should be ample, but you can always buy more hours if necessary, and don't be afraid to ask. I did.

The part 2 test takes about an hour. It's based on the normal driving test, but the examiners will expect to see a much higher standard of driving. You will pass if you have no serious or dangerous faults and a maximum of 6 driving faults.

Not all driving test centres offer part 2 tests, so check online where your nearest part 2 test centre is and ideally try and familiarise yourself with the local roads.

You're allowed a maximum of 3 attempts to pass your part 2 test. If you fail all 3 attempts then you have to wait for 2 years after you passed your theory and start over again (or find another career!)

So, on to phase 2...your **ADI Part 3 training**.

This is where the fun really starts.

You have two options open to you here, I've named them phase 2a) and phase 2b).

Which option you decide to go down is down to your own circumstances and what your training company can provide.

Phase 2a)

This entails completing 40 hours of training and then taking your ADI part 3 exam.

At any time in your 40 hours of training, you can change your mind and apply for a PDI licence instead of taking your ADI part 3 at the end. Once you've started your part 3 training, you'll soon know which option is right for you.

So arrange your 40 hours of training and book your part 3 exam on the .gov.uk website to coincide with the end of the training. The cost of the test is £111.

Don't wait until you complete your 40 hours of training before going online to book your test, as the waiting time can be as long as 3 months in some parts of the country.

If you're successful on your first attempt at your part 3 test, then you'll join a very elite group. The pass rate is VERY low, so beware.

If you're not successful on your first attempt, then you can have a further two attempts (at a cost of £111 each).

If you fail all three attempts at your part 3 test then you have to wait for 2 years from the date you passed your part 1 theory test and start the whole process over again.

Once you've passed your ADI part 3 test you can then apply to the DVSA for your first ADI certificate (your green badge). Registering with the DVSA costs £300. You have to renew your registration every 4 years, which costs a further £300 each time.

Phase 2b)

Once you've completed your 40 hours of part 3 training, you can apply for a trainee (PDI) licence. This costs £140 and is valid for 6 months. This not only enables you to charge pupils for driving lessons but it also helps you gain valuable experience instructing pupils to drive, so that you can prepare thoroughly for your part 3 test.

You have 2 options to choose from when you apply for a trainee PDI licence:

Option a) be supervised for 20% of all lessons you give whilst you have a PDI licence, or

Option b) do at least 20 further hours of training while you have the PDI licence.

You must complete the 20 hours of training before you book your ADI part 3 test and you have to book your test within 3 months of getting your PDI licence. At least 5 hours of the training must be practical in-car training, whilst the other 15 hours can be either in-car or classroom-based, or a combination of the two.

You can only choose one of the above options and you can't change to the other one once you've made your decision. You'll need to talk with your training organisation about which option is best for you. Most instructors choose option b).

In my own part 3 training, I chose to go down the PDI route and I'm very glad I did. The experience I gained with real pupils was invaluable (and it gave me an income as well). I chose option b) and received a further 20 hours of training, all of which I did in-car (at considerable extra cost, but which was worth every penny).

My trainer for these 20 hours was the former Head of Examiner Training at the DVSA Headquarters, and it was his high-quality training that got me through my ADI part 3 test on my third and final attempt with a near-perfect score. I mention this not to brag, but to emphasise how important the quality of training is. Had I remained with my original instructor, I am certain that I would not have passed.

Whether you decide to go down the PDI route before taking your part 3 test is ultimately your decision.

The imperative thing is that you check that your training organisation offers both options. If they don't, then I suggest you look elsewhere.

CHAPTER 3 –

To Franchise or not to Franchise? ...that is the question

Most existing ADI's and PDI's will have thought about joining a franchised driving school at some point in their driving career, usually at the start.

So let's have a balanced look at the main pro's and con's of both. I've taken the best case scenario for the pro's and the worst case scenario for the con's.

Clearly, what's included in a franchise differs between franchisors, but this should give you a flavour.

FRANCHISE - The Pro's

- Regular supply of new pupils
- Your diary can be filled in a short space of time
- Fixed weekly or monthly franchise fee
- Car provided (serviced and maintained)
- Support helpdesk
- On-going training
- Business stationery
- Corporate branding
- No marketing costs
- Lesson fees collected for you
- Your diary managed for you.

FRANCHISE - The Con's

- Franchises are usually expensive, often in excess of £250 per week, meaning you can be working long hours for little or no profit
- Contractual tie-ins, usually 12 or 24 months, sometimes with no break clauses
- Your part 2 & part 3 trainer is allocated to you. You usually have no choice. He/she may be really good, or not
- Pupils allocated to you may be miles apart, costing you fuel and lost time
- Admin fees are often deducted from lesson income, so you may not receive the full amount that the pupils have paid for lessons

- You must adhere to the school's terms and conditions
- Limited opportunity to develop your own business ideas
- Limited choice of car
- You may still have to pay your franchise fee when you take time off for holiday or sickness.

INDEPENDENT - The Pro's

- You're completely your own boss
- Total freedom to work your own hours and charge your own rates
- No franchise fee to pay....lower overheads mean higher profits
- Choose your own car - any make, new or used
- Full control of which pupils you take on and their locations
- Develop your business in whichever way you choose.

INDEPENDENT - The Con's

- Hard work initially. Its success is down to you
- Requires a degree of business acumen and strong interpersonal skills
- Needs sound planning and diary management skills
- Can feel like you're on your own
- Need to provide and fund your own driving car
- Need to spend time and money on marketing and admin.

There's no right or wrong answer as to which route you decide to go down. It's whatever works best for you.

I've spoken with many other driving instructors on this subject. Most of them (including myself) started their driving career in some form of conjunction with an existing driving school. This could be either a national franchise or with a local driving school. Not many driving instructors I know started their careers as fully independents. Most got some form of help in their early days.

My advice is not to rush into anything or make a hasty decision. Remember that you're still self-employed irrespective or whether or not you go down the franchise route. Therefore you need to be developing your business in a way that works for you.

A question I'm often asked is:

If I was starting out again now, what would I do now, given what I know about the industry?

That's a really good question. Here are my thoughts:

- I would **keep my options open** for as long as I could. I would certainly not commit to a long-term contract before I had started my training
- I would also want to be in control of choosing my trainer, as opposed to having to stick with the one allocated to me (I actually changed my trainer three times during my training. Had I not done so, I am certain that would not have passed my ADI part 3 test)
- I would find out as much as I can about my local area by talking with other local driving instructors (this worked brilliantly for one of my lady driving instructor pupils…see her story below)
- I would then make an **INFORMED** decision, having weighed up the pro's and con's (and costs) of going independent or joining a franchise.

What worked for me when I started out was joining forces with the driving school that was top of the Google rankings for "driving lessons Hastings". I paid a weekly fee of £50 and in return I had an abundant supply of local pupils. My diary was full within two weeks of joining them, and most importantly I was not tied in to any long-term agreement. Instead I could leave at any time by giving just one month's notice.

That's what worked for me and it's certainly worth considering.

To check it out in your own area, just do a Google search on "driving lessons Hull", or "driving lessons Mansfield" or wherever you live. Then scroll down the search results and ignore any large nationals and any paid advertising. Contact the driving schools which come in the top 3 of the organic search and see what they have to offer. Quite often you'll get many of the benefits of a large driving school at significantly lower cost and without any long-term contractual tie-in. Most local driving schools have flexible terms, so you may be able to negotiate your own deal with them.

Case Studies

Case Study 1. Ian

Ian was formerly a retail manager working in a major city centre retail outlet. He had to work most Saturdays and often on Sundays as well. This meant long hours at work, plus up to an hour's commute each way every day. He had two young children at primary school age. Ian's major "why" he wanted to become a driving instructor was because he wanted to spend more time at home with his family.

Ian trained with me for his ADI part 2 and part 3 tests. He decided to do his part 3 test at the end of his part 3 training, as he wanted to qualify as an ADI before handing in his notice at work. This was definitely the right decision for Ian, as there is no guarantee that you will pass your part 3 test and Ian had a young family to support.

Ian is a very organised guy who pays close attention to detail. He soaked up the training and passed his part 3 test on his second attempt. He did extensive research into driving instructor franchises and joined one of the smaller, less-well-known national franchisors.

He gets a regular supply of pupils and he negotiated a break clause in his contract after 6 months if he wasn't satisfied with the level of marketing support he was promised. He also negotiated only paying for 48 weeks every year, meaning that he didn't have to pay his weekly franchise fee when he took his main holidays.

I have kept in touch with Ian. He loves his new career and it's worked out really well for him. He recently told me "I've got my life back. It's so nice having weekends off. I can be a proper dad now".

Case Study 2 : Rebecca

Rebecca was allocated to me from one of the major national driving instructor training schools (I cover Kent and East Sussex for them). I was her trainer for her ADI part 2 and part 3 tests. Rebecca had recently lost her job and needed to replace her income as quickly as she could.

She passed her ADI Part 2 test on her first attempt (with only 1 minor driving fault) and she then did an intensive course with me for her 40 hours of ADI part 3 training. At the end of this training she obtained a PDI licence and joined a local driving school. Her diary was full within 3 weeks of joining them, but she found that the quality of pupils allocated to her wasn't that great, because of their introductory offer of 'first 2 hours of lessons for £10'. This attracted several pupils who simply wanted to get a couple of hours of cheap lessons, before moving on to the next introductory offer.

Whilst waiting at the local test centre for pupils to come back from their driving test, Rebecca used to make a point of speaking with the other instructors. She discovered that there were no female instructors in the area offering lessons in an automatic car. This was her niche that she had discovered simply by asking around. And that's the route she went down after qualifying. She took her first ADI part 3 test after 4 months on her PDI licence and narrowly failed. She passed on her second attempt six weeks later.

Rebecca is now a completely independent driving instructor offering lessons in her automatic car. She gets all her new pupils from either word of mouth or from local advertising and she normally has a 6 to 8 week waiting list. She works on a part-time basis, which she can afford to do, as she doesn't have the financial burden of weekly franchise fees. Nor does she have the cost of an additional car, as she adapted her own car for driving lessons simply by having dual controls fitted.

CHAPTER 4 – The ADI Driving Instructor Tests Explained

There are 3 tests you have to pass in order to qualify as an approved driving instructor. These are…

- ADI Part 1 Theory (part 1)
- ADI Part 2 Practical Driving (part 2)
- ADI Part 3 Teaching (part 3)

Once you've passed your Part 1 test, you have 2 years in which to pass parts 2 and 3.

So let's have a detailed look at what's involved in each test…

ADI Part 1 (Theory)

There are 2 sections to the Part 1 Test

- 100 multiple choice questions, and
- 14 hazard perception video clips

Section 1 – Multiple Choice

You get 90 minutes to answer 100 questions. There are 4 possible answers for each question and you select your answer by either touching the computer screen or by using a mouse. You need to score at least 85/100 in order to pass the theory part of the test.

The multiple-choice section is banded into four main areas, each with 25 questions…

- Road procedure
- Traffic signs, traffic signals, car control, pedestrians, mechanical knowledge
- Driving Test, disabilities, law
- Publications, instructional techniques.

Although the overall pass is 85%, you must reach a minimum mark of 80% in each of the 4 subject areas. So you need at least 20 out of 25 correct in each of the four bands. For example a score of 25, 25, 25 and 19 would be 94% but would be a fail.

A sample question is....

Your vehicle has broken down on a motorway. You are not able to stop on the hard shoulder. What should you do? (Choose 1)

- *Switch on your hazard warning lights*
- *Stop following traffic and ask for help*
- *Attempt to repair your vehicle quickly*
- *Stand behind your vehicle to warn others*

Section 2 - Hazard Perception

- You're shown 14 video clips, one after the other
- Thirteen clips contain one hazard and one clip contains two hazards, so there are 15 hazards in total. You score up to 5 points on each hazard, depending how quickly and correctly you react
- The maximum score is 15 x 5 = 75 points
- You have to get at least 57 points to pass.

From my own experience, there is plenty of time to answer all the multiple-choice questions and to go back and check them. I was certain of the answers to 94 of the 100 questions, but I wasn't sure about the other remaining 6, so I made my best guess and scored a total of 97. A decent pass.

I found the hazard perception part trickier, as the videos keep coming at you in quick succession, and you have to keep concentrating.

On one of the clips, I clicked too early and scored 0!! But I still passed this part OK with 64/75.

The key to passing your ADI part 1 first time is practice. There are loads of books, apps, DVD's and online resources to help you.

The one I found most useful is www.drivingtheory4all.co.uk

Driving Theory 4 All

https://www.drivingtheory4all.co.uk/

Theory test practice material for every DVSA **theory test**. Driving Theory **4 All** is the smart, fast and personal way to revise for your driving **theory test**.

It's well worth the relatively small cost and it covers both the multiple choice and the hazard perception.

The ADI Part 1 test shouldn't cause you any problems so long as you've practiced sufficiently. If you're not successful then you can retake this test as many times as you need to.

ADI Part 2 (Practical Driving Test)

Please download the practical driving test marking sheet (it's known as form DL25A).

To do so, enter **'DL25A'** in a Google search, download it and print it off.

The ADI part 2 test uses the same marking sheet as is used for normal driving tests (the 2017 version of the marking sheet is shown below), but the standards expected are much higher.

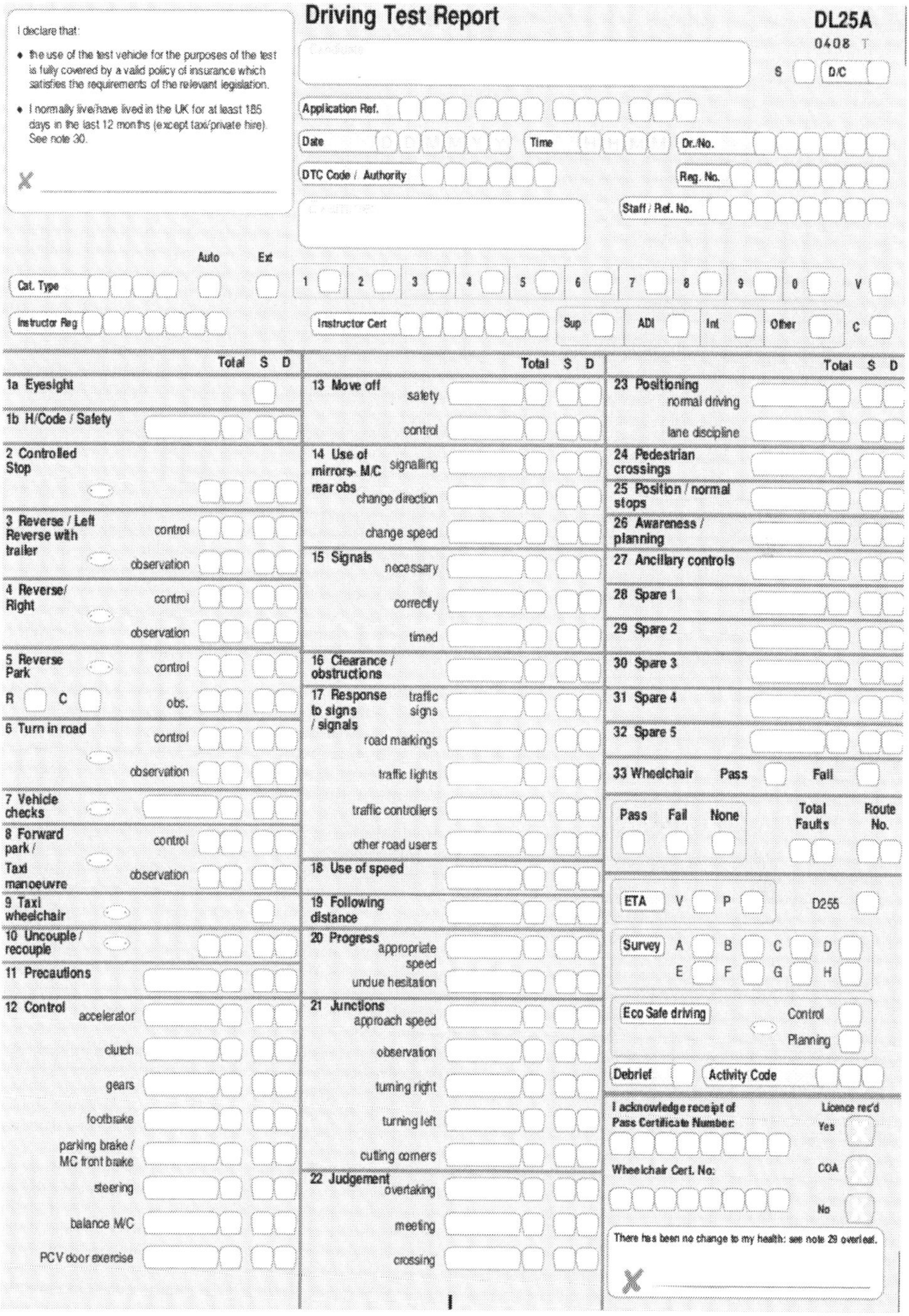

I declare that:

- the use of the test vehicle for the purposes of the test is fully covered by a valid policy of insurance which satisfies the requirements of the relevant legislation.
- I normally live/have lived in the UK for at least 185 days in the last 12 months (except taxi/private hire). See note 30.

X

Driving Test Report **DL25A**

0408 T

Candidate

S D/C

Application Ref.

Date D D M M Y Y Time H H M M Dr./No.

DTC Code / Authority Reg. No.

Examiner Staff / Ref. No.

Cat. Type Auto Ext 1 2 3 4 5 6 7 8 9 0 V

Instructor Reg Instructor Cert Sup ADI Int Other C

	Total	S	D
1a Eyesight			
1b H/Code / Safety			
2 Controlled Stop			
3 Reverse / Left Reverse with trailer – control			
observation			
4 Reverse/ Right – control			
observation			
5 Reverse Park – control			
R C – obs.			
6 Turn in road – control			
observation			
7 Vehicle checks			
8 Forward park / Taxi manoeuvre – control			
observation			
9 Taxi wheelchair			
10 Uncouple / recouple			
11 Precautions			
12 Control – accelerator			
clutch			
gears			
footbrake			
parking brake / MC front brake			
steering			
balance M/C			
PCV door exercise			

	Total	S	D
13 Move off – safety			
control			
14 Use of mirrors- M/C rear obs – signalling			
change direction			
change speed			
15 Signals – necessary			
correctly			
timed			
16 Clearance / obstructions			
17 Response to signs / signals – traffic signs			
road markings			
traffic lights			
traffic controllers			
other road users			
18 Use of speed			
19 Following distance			
20 Progress – appropriate speed			
undue hesitation			
21 Junctions – approach speed			
observation			
turning right			
turning left			
cutting corners			
22 Judgement – overtaking			
meeting			
crossing			

	Total	S	D
23 Positioning – normal driving			
lane discipline			
24 Pedestrian crossings			
25 Position / normal stops			
26 Awareness / planning			
27 Ancillary controls			
28 Spare 1			
29 Spare 2			
30 Spare 3			
31 Spare 4			
32 Spare 5			

33 Wheelchair Pass Fail

Pass Fail None Total Faults Route No.

ETA V P D255

Survey A B C D E F G H

Eco Safe driving Control Planning

Debrief Activity Code

I acknowledge receipt of Pass Certificate Number: Licence rec'd Yes

Wheelchair Cert. No: COA

No

There has been no change to my health: see note 29 overleaf.

X

Show me, tell me questions...

You will be asked 3 'tell me' questions (where you explain how you'd carry out a task) at the start of your test, before you start driving

You also be asked 2 'show me' questions while you're driving.

The **'tell me'** questions you could be asked are:

1. Tell me how you'd check that the brakes are working before starting a journey.

The brakes should not feel spongy or slack. Brakes should be tested as you set off. The car shouldn't pull to one side when you're braking.

2. Tell me where you'd find the information for the recommended tyre pressures for this car and how tyre pressures should be checked.

In the car manual, use a reliable pressure gauge when the tyres are cold.

3. Tell me how you make sure your head restraint is correctly adjusted.

The middle of the head restraint should be in line with your eyes or the top of your ears.

4. Tell me how you'd check the tyres to ensure that they have sufficient tread depth and that their general condition is safe to use on the road.

No cuts and bulges, with at least 1.6mm of tread depth across the central three-quarters of the breadth of the tyre, and around the entire outer circumference of the tyre.

5. Tell me how you'd check that the headlights and tail-lights are working.

Explain you'd turn the lights on (turn on ignition if necessary), then say you'd walk round vehicle and check they're working ok.

6. Tell me how you'd know if there was a problem with your anti-lock braking system (ABS).

A warning light should illuminate on the dashboard if there's a fault.

7. Tell me how you'd check the direction indicators are working.

Explain that you'd operate the indicator switch (turn on ignition if necessary), and then walk round vehicle. Alternatively, say you'd turn the hazard lights on and walk round the vehicle.

8. Tell me how you'd check the brake lights are working.

Explain you'd press the brake pedal, make use of reflections in windows or doors, or ask someone to walk round the back and help.

9. Tell me how you'd check the power-assisted steering is working before starting a journey.

If the steering becomes heavy, the system may not be working properly.

10. Tell me how you'd switch on the rear fog light(s) and explain when you'd use it/them.

Turn the rear fog lights on (turn on dipped headlights and ignition if necessary). Check the warning light is on. Fog lights should be used when visibility is less than 100 metres.

11. Tell me how you switch your headlight from dipped to main beam and explain how you'd know the main beam is on.

Operate the headlight switch and check the main beam light is on (usually blue).

12. Open the bonnet and tell me how you'd check that the engine has sufficient oil.

Identify the dipstick. Say you'd pull it out, wipe it, then put it back in. Then you'd pull it out and check the oil level is between min and max.

13. Open the bonnet and tell me how you'd check that the engine has sufficient engine coolant.

Identify the engine coolant reservoir and say that the engine coolant level should be between min and max.

14. Open the bonnet and tell me how you'd check that you have a safe level of hydraulic brake fluid.

Identify the brake fluid reservoir, then check that the level is between the min and max markings.

You'll also be asked two of the following **'Show me'** questions while you're driving....

When it's safe to do so...

1. Show me how you wash and clean the rear windscreen.
2. Show me how you wash and clean the front windscreen.
3. Show me how you'd switch on your dipped headlights.
4. Show me how you'd turn on the rear demister.
5. Show me how you'd operate the horn.
6. Show me how you'd demist the front windscreen.
7. Show me how you'd open and close the side window.
8. Show me how you'd operate the cruise control.

The standards expected are quite high, so you can't treat this test as a stroll in the park, just because you're a full licence holder.

I give my instructor pupils ten hours of in-car training and I always start by saying this to them: "I'm not here to teach you how to drive. You're a full licence holder and you've been driving for many years and I fully respect that. My job is purely to get you through your part 2 test and nothing more. I've sat in the back of many part 2 tests and I know exactly what the examiners are looking for. Is that ok?"

All of my pupils have all replied "thanks, that's exactly what I want" (or words to that effect).

Many of my pupils take their test in Eastbourne, which is the main area I teach ADI part 2. But others who have trained with me have taken (and passed) their tests in locations all over the UK including Birmingham, Exeter and Aberdeen! It doesn't make that much difference, so long as you use the disciplines you've learnt in the training.

If you use your own car then you need to supply a second interior mirror. You can buy these for under £5 at Halfords. If you forget, most test centres do have their own, but bring your own spare mirror just to be on the safe side.

The test takes around an hour and will include dual carriageways (including motorways), 'A' roads, town centre driving, residential areas and country lanes. There will be around 20 minutes of independent driving where you will either follow satnav or the examiner will tell you to follow signs for town centre, ring road etc.

Going the wrong way is NOT a test fail, so long as you drive safely. At worse it's a driving fault for 'planning and awareness'. If in doubt you can ask the examiner for clarification. The test will include an emergency stop and two of the following reversing manoeuvres...

- Forward bay park
- Reverse bay park
- Parallel park
- Stop on the right, reverse two car lengths and re-join the road safely.

To give you an idea of the sort of standard the examiners are looking for, my pupils have been given faults for the following...

- Driving at only 62mph on a dual carriageway, when it was safe to do 70mph
- Driving at 23mph in a 30mph area

- Not overtaking a bus when it was safe to do so
- Stopping too close to the car in front (tyres and tarmac rule) in queueing traffic
- Not leaving sufficient clearance when passing a parked car
- Looking over your right shoulder when joining a motorway
- Not signalling left to switch lanes after overtaking a lorry on a dual carriageway
- Not giving way when an oncoming vehicle had priority
- Incorrect road position (driving too close to the centre line of the road)
- Looking in the door mirrors too much when reversing
- Mirror checks done too late when switching lanes.

These are just examples of the types of faults that have caught some of my instructor pupils out.

Without tempting fate, all my pupils have passed their part 2 test within the permitted three attempts. The majority pass first time and the remainder on their second attempt. Only one of my pupils took three attempts. I've found that 10 hours of in-car training is ample. We cover all the reversing manoeuvres and we also go through the marking sheet line by line, so that the pupil knows the standard that is expected at whichever test centre the pupil is taking their test.

I also train my pupils to use 'commentary driving' when appropriate. I've found that the examiners welcome this. For example "I had to cut that corner slightly because of the parked car". It's a really good technique to use in a part 2 test.

Get yourself a good instructor, prepare well for the test and you should be fine.

Then the fun bit starts: onto your part 3 training!

ADI Part 3 (Instruction Ability)

This is by far the most difficult of the three tests to pass. The overall pass rate in 2017 was a mere 27%, so I'm going to go through this test in detail.

The new ADI part 3 test was introduced in December 2017 and is based on the same standards check test that all ADI's have to take every 2 to 4 years to ensure that teaching standards are kept to a high standard.

The test involves you (the instructor) conducting a lesson of your choice for a pupil of your choice. You sit in the front passenger's seat and your pupil drives. The examiner sits in the back, observes your lesson and marks the score sheet.

The whole test takes around 60 minutes from start to finish, including a briefing to your pupil, the lesson itself and a recap at the end.

There are two core elements that you must demonstrate throughout your lesson.

The first of these is called "**Core Competencies**". By this I mean...

- Fault Identification
- Fault Analysis
- Remedial Action.

Your teaching needs to incorporate all three of these core elements.

Using an example of a driver approaching a T-junction too fast... what you WOULD NOT simply say is "slow down" (remedial action) or "you're driving too fast" (fault identification only).

What you would say is...

"Slow down to 10mph" (control the situation first). "Turn left at the junction and then stop at a safe place on the left".

When the pupil has stopped, go through the 3 core competencies in order...

- Your approach speed was too fast (**fault identification**)
- The danger of approaching too fast is that it might cause you to stop beyond the give-way line at the junction and maybe hit a cyclist or a car (**analysis)**
- So on approach to the junction, slow down to 10mph three car lengths before you reach the junction (**remedial action**).

The second core element is that you must demonstrate that learning has taken place. This is demonstrated by your **teaching style**, of which there are three types, namely…

- Full instruction (eg check your centre mirror, now your left door mirror)
- Prompted practice (eg which mirrors are you going to check?)
- Independent driving (give the pupil feedback after the junction)

So learning can be deemed to have taken place (as an example) by starting your lesson off using full instruction, then when you're happy that the pupil has done a certain element correctly two or three times, you start to move your teaching style on to prompted practice for certain elements (eg checking mirrors), as opposed to giving full instruction all the time.

ADI Part 3 (SC)

Driver & Vehicle Standards Agency

INFORMATION

Candidate's name		Location		Outcome
PRN ☐☐☐☐☐☐		Date ☐☐/☐☐/☐☐		

I declare that my use of the test vehicle for the purposes of the test is covered by a valid policy of insurance which satisfies the requirements of the relevant legislation.

Dual Controls Yes ☐ No ☐ Reg No ☐

Log book Yes ☐ No ☐ Trainer PRN ☐☐☐☐☐☐

Trainee Licence Yes ☐ No ☐ ORDIT Yes ☐ No ☐

Accompanied? QA ☐ Trainer ☐ Other ☐

ASSESSMENT

Lesson

Student: Beginner ☐ Partly trained ☐ Trained ☐ FLH New ☐ FLH Experienced ☐

Lesson theme:

Junctions ☐ Town & city driving ☐ Interaction with other road users ☐
Dual carriageway / faster moving roads ☐ Defensive Driving ☐ Effective use of Mirrors ☐
Independent driving ☐ Rural roads ☐ Motorways ☐ Eco-safe driving ☐
Recap a manoeuvre ☐ Commentary ☐ Recap Emergency stop ☐ Other ☐

		Competence 0	1	2	3
		No evidence	Demonstrated in a few elements	Demonstrated in most elements	Demonstrated in all elements
Lesson Planning					
Did the trainer identify the pupil's learning goals and needs?					
Was the agreed lesson structure appropriate for the pupil's experience and ability?					
Were the practice areas suitable?					
Was the lesson plan adapted, when appropriate, to help the pupil work towards their learning goals?					
Score for lesson planning	☐				
Risk Management					
Did the trainer ensure that the pupil fully understood how the responsibility for risk would be shared?					
Were directions and instructions given to the pupil clear and given in good time?					
Was the trainer aware of the surroundings and the pupil's actions?					
Was any verbal or physical intervention by the trainer timely and appropriate?					
Was sufficient feedback given to help the pupil understand any potential safety critical incidents?					
Score for risk management	☐				
Teaching & learning strategies					
Was the teaching style suited to the pupil's learning style and current ability?					
Was the pupil encouraged to analyse problems and take responsibility for their learning?					
Were opportunities and examples used to clarify learning outcomes?					
Was the technical information given comprehensive, appropriate and accurate?					
Was the pupil given appropriate and timely feedback during the session?					
Were the pupils queries followed up and answered?					
Did the trainer maintain an appropriate non-discriminatory manner throughout the session?					
At the end of the session – was the pupil encouraged to reflect on their own performance?					
Score for teaching and learning strategies	☐				
Overall score	☐				

REVIEW	YES	NO
Did the candidate score 7 or less on Risk Management (A 'Yes' response to this question will result in an automatic Fail)		
At any point in the lesson, did the candidate behave in a way which put you, the pupil or any third party in immediate danger, so that you had to stop the lesson (A 'Yes' response to this question will result in an automatic Fail)		
Was advice given to seek further development?		

Feedback offered to Candidate			
Examiner Name		Signature	

[illegible]/2017

The score sheet (above) can be downloaded by doing a Google search “ADI part 3 scoring sheet”. Here’s the search result…

About 12,000,000 results (0.41 seconds)

Search Results

Web results

Approved driving instructor (ADI) part 3 test: marking sheet - GOV.UK

https://www.gov.uk/government/.../approved-driving-instructor-adi-part-3-test-report

1 Sep 2017 - The report **form** used to mark your approved driving instructor (**ADI**) **part 3 test**, which assesses your ability to teach pupils.

Download it and print two copies off if possible (one to write notes all over and the other to keep for reference).

Choosing your pupil

It’s really important that you choose the right pupil. Your pupil can be of any standard, including:

- Beginner
- Partly trained (typically 6 to 15 hours driving experience)
- Trained (15 to 30 hours, but not passed their test yet)
- Full licence holder (FLH) New (just passed their test)
- Full licence holder (FLH) Experienced (passed their test several years ago, but has picked up some bad habits).

You choose your lesson theme, which MUST be appropriate for your pupil. Examples of lesson themes are:

- Moving off and stopping
- Junctions (left and right junctions and T junctions)
- Crossroads
- Town/city centre driving
- Roundabouts
- Meeting oncoming traffic
- Dual carriageway / faster roads
- Pedestrian crossings
- Rural roads
- Reverse parallel parking
- Bay parking (forwards and reverse)
- Emergency stop

Your lesson theme must match the needs of the pupil. So, for example, you wouldn't do reverse parallel parking with a beginner, nor would you do moving off and stopping with a full licence holder.

The lesson theme must also meet a real need. Examiners are trained to see through a lesson where the pupil has been briefed to make pre-determined faults (this is known as staging). Don't do it. Find a pupil with a genuine weakness or need and design a lesson accordingly.

There are examples of lesson themes on the test report sheet, but it doesn't have to be any of these (see the box marked "other"). The theme of the lesson can be anything of your choice, so long as it's relevant to the pupil's needs.

Examples of lesson themes that have worked well for my pupils in the past include:

- Moving off and stopping for a Beginner
- Junctions for a Beginner. In such a lesson, find a quiet residential area that you can go round doing just left turns. Then turn the car round and do right turns.
- Roundabouts for Trained or FLH New driver
- Approaching junctions for FLH New or FLH Experienced. The pupil was an experienced driver who approached junctions too fast
- Meeting oncoming traffic for a Trained driver. When to give way, when not to. Demonstrating hold-back positions and anticipation of other road users
- Reverse bay parking for FLH who had a complete fear of reversing. In this lesson, we started in a quiet area of a large supermarket car park, worked on the technique and then executed it for real in between parked car in a busier part of the car park
- Overtaking on fast roads for FLH (new) who had limited experience of driving on dual carriageways and who was nervous about overtaking
- Driving along country lanes for Trained driver. Appropriate speeds, correct gear, road positioning, anticipation, meeting and passing oncoming traffic.

Obviously you can't do rural roads or motorway driving if there isn't a suitable area within easy reach of the test centre. What works well for one test centre may not work at another test centre.

Choose the right pupil and the right lesson theme and you're well on your way.

I've spoken with a number of ADI part 3 examiners about what advice they can offer pupils preparing for their ADI part 3 test. Their feedback is as follows:

"Take great care with your choice of pupil. Find out what their needs are and then plan a lesson to address these needs".

So don't do it the other way round, i.e. don't pick a lesson theme and then try and get a pupil to fit it...the examiners can see right through it.

"Make sure the route you choose is appropriate".

In other words, if your lesson theme is roundabouts, make sure your route has a sufficient number of roundabouts of different sizes and complexity, even if this means approaching the same roundabout from different directions.

"Make sure that you have progress options for your pupil".

For example, if you're teaching reverse bay parking in a quiet car park and your pupil does it correctly on their third attempt, then you need to be able to develop the lesson. So be prepared to drive to a busy supermarket and continue the lesson theme there.

I know of an instructor whose lesson theme in his ADI part 3 test was parallel parking. Unfortunately his pupil did the manoeuvre correctly on her first attempt. They then drove to a downhill road, where the pupil successfully did an uphill parallel park on her first attempt. There was still 30 minutes of the test to go and the trainee instructor had not prepared anything further. As a result, he failed. So be warned. You must have progress options.

"Adapt the lesson plan where necessary".

For example, if your lesson theme was pedestrian crossings and you noticed that your pupil was approaching junctions too fast, then change the lesson plan and work on junctions instead. Don't stick rigidly to your original plan.

"Make sure that your level of instruction is appropriate".

There are three levels of instruction, namely full instruction, prompted practice and independent driving. Make sure that the level of instruction is appropriate and vary it when you feel the pupil is ready. For example, for a beginner you may say "check your centre mirror now and your left door mirror. Now signal left." Once they had done this correctly 3 or 4 times, you would consider changing your level of instruction to prompted practice and say "which mirrors are you going to check? When should you signal?" Don't hold the pupil back unnecessarily, nor jump to the next level before they're ready.

"Explain the consequences of any serious faults they make".

For example, don't just say "you forgot to check your blind spot". Instead say "you forgot to check your blind spot [fault indentification]. The danger of not checking your blind spot is that you could pull out and hit a car that was passing you [fault analysis]. So before you move off from the side of the road, do your observations starting from the safe side round to the danger side, finishing off with a blind spot check [remedial action]".

The Scoring Sheet

The scoring sheet is split into 3 sections, namely:

Section 1 **Lesson Planning** (4 scoring areas, each scoring 0 – 3 points)

Section 2 **Risk Management** (5 scoring areas, each 0 – 3 points)

Section 3 **Teaching and Learning Strategies** (8 scoring areas, each 0 – 3 points).

So there are 17 scoring areas, each scored 0, 1, 2 or 3.

- A mark of 0 is given if there is no evidence
- A mark of 1 is awarded if the relevant skill is demonstrated in a few areas
- A mark of 2 is awarded if the relevant skill is demonstrated in most elements
- A mark of 3 is awarded if the relevant skill is demonstrated in all elements.

The maximum score is 17 x 3 = 51. There are 3 possible outcomes for the test, namely:

Score	Description	Grade
0 – 30	Unsatisfactory performance	FAIL
31 – 42	Sufficient competence demonstrated to permit entry to the Register of Approved Driving Instructors	GRADE B
43 – 51	A high overall standard of instruction demonstrated	GRADE A

So your target is to score 31 points or more out of 51.

Let's have a detailed look at each of the 17 skills and what you have to do in order to get top marks in each. Let's explore them section by section:

Section 1 Lesson Planning

1. Did the trainer identify the pupil's learning goals and needs?

This takes place right at the start of the test. You need to rehearse this with your pupil and get him/her to tell you what they want to achieve from the lesson, or alternatively (what I recommend you do) is to prompt your pupil.

For example, you may have a quick conversation with your pupil (Jane) along the following lines:

"Ok Jane, let's have a quick recap on your last lesson. Can you remember what we covered?"

"I moved off from the side of the road and then drove for a bit in first gear and then you got me to stop on the left".

"How did you feel this went?"

"Alright. I was a bit unsure at first when to press the clutch down, but I did it quite a few times OK without you having to tell me what to do".

"Yep, you certainly did. So what would you like to achieve today?"

"I'd like to build on what we did last time and learn how to turn left and right at junctions".

"OK, when you say junctions, do you turning left and right or T-junctions?"

"Well, both really".

So, you've listened carefully to your pupil and identified what they want to achieve. You've asked relevant questions and not made any assumptions.

Verdict: This should be an easy 3 points.

2. Was the agreed lesson structure appropriate for the pupil's experience and ability?

The lesson structure should allow the pupil to be stretched, but not overwhelmed. In other words, it should be at or slightly above their level of competence and they should be able to show progress during the lesson. For example, you wouldn't take a beginner into a busy town centre, not would you restrict a pupil approaching test standard to turning left and right in a quiet residential area.

The keys to getting a high mark in this section are:

- Making sure that the lesson structure meets the pupil's needs (as identified in point 1)
- Making sure that the pupil understands what they'll be doing in the lesson and agrees with it
- Making sure that there are opportunities for progressing, as and when you're satisfied that the pupil has achieved what was set out and agreed.

Avoid...

- Giving a standard lesson that's not tailored to meet the needs of the pupil.

Verdict: This needs careful planning. With care, you should be looking to score at least 2 points.

3. Were the practice areas suitable?

The route that you choose needs to be appropriate to meet the needs of the pupil and achieve the agreed lesson structure.

So you need to make sure that the route will offer a sufficient number of challenges that the pupil can rise to, but not overwhelm them. In other words, if they're level of driving ability is 5 (on a scale of 1 to 10, 10 being test-ready), then the route should have sections at levels 5, 6 and 7.

Ideally, you need to think of contingencies…if the pupil manages the agreed lesson structure on the selected route/area easily then drive to a busier area. Conversely, if the pupil is struggling, then drive to a quieter area.

Verdict: the route and practice area needs careful planning in conjunction with your trainer. This is where your trainer earns his/her crust by helping you choose the right route to match the needs and driving ability of your pupil.

A score of 2 here is OK.

4. Was the lesson plan adapted, when appropriate, to help the pupil work towards their learning goals?

This can be a tough one and it requires you to assess what is going on and decide whether you need to change the lesson plan based on how your pupil is performing on the day. You need to be on your game and you may need to make the lesson more challenging if the pupil is not being stretched sufficiently. Alternatively, you may need to simplify the lesson structure if you feel your pupil is struggling on the day.

Remember that your pupil may be feeling nervous, even though they're not to one being assessed. They may feel uncomfortable on the day with someone sitting in the back observing them. You need to pick up on this and be ready to adapt the lesson plan if needs be.

A good tip is to get your pupil to pull over on the left during the lesson and ask them how they're feeling. If it's not going quite to plan, don't panic, as you'll score well here by showing that you've identified that the lesson plan needs adapting and that you're taking appropriate action. You should discuss this with your pupil and explain why you're making changes to the agreed lesson plan.

Avoid…

- Ignoring any issues or faults your pupil is making and continuing on regardless with the agreed lesson plan
- Not telling your pupil that you're changing the lesson plan nor explaining why.

Verdict: This is a tough one, as you can't plan it in advance. You need to be highly observant and make changes if necessary.

Be happy with a 2.

Section 2 Risk Management

Risk management falls into two very separate areas:

Firstly, you are responsible for the safety of other road users including pedestrians throughout the lesson. This means using the dual controls (if your car is fitted with them) and taking control of the steering wheel if necessary.

Secondly, it means that your pupil must be fully aware of how the risk is going to be shared between you and him/her. Exactly how the risk is shared will depend on how competent your pupil is.

There are five scoring areas in this section, namely:

5. Did the trainer ensure that the pupil fully understood how the responsibility for risk would be shared?

There are two main parts to this, namely

- An opening statement on how the risk is going to be shared, and
- Varying the amount of responsibility for risk the pupil has during the lesson. This normally means giving them a greater share of the risk as they progress

Examples of opening statements are:

a) For a beginner pupil:

"I expect you to control the car and drive safely with due consideration for our safety and for the safety for other road users including pedestrians at all times. I have dual controls of clutch and brake, so I can take control of the car if I need to. I may also need to take hold of the steering wheel. If I do so, please relax your grip on the steering wheel. If I need to take control I will tell you so and explain why". Please don't worry about this, as it's quite normal for new drivers.

b) For a full licence holder:

"As a full licence holder I expect you to take full responsibility for our safety and for the safety of other road users including pedestrians. I will try and keep my comments to a minimum and I'll only say anything if I feel you've missed something important. So if I'm quiet, take that as me being comfortable with your driving".

The second area of risk management is to vary the amount of responsibility the pupil has when it's appropriate to do so. For example, you may be practicing turning left at junctions and your pupil has checked their mirrors to your satisfaction 3 or 4 times in a row with you prompting them to do so. So you may say:

"Well done on your mirror checks when approaching junctions. I'm going to hand this responsibility over to you now, so I expect you to continue to check your mirrors exactly as you have been doing, but without me prompting you. I will be watching to make sure you check your mirrors, but I won't make any comments unless I need to. Is this OK with you?"

Avoid:

- Forgetting to explain how responsibility for risk will be shared at the beginning of the lesson
- Not handing over more responsibility for risk to the pupil when they were capable of taking it.

Verdict: your opening statement should be prepared in advance and you can refer to written notes if you need to, but avoid just reading it out aloud parrot fashion. Remember to hand at least one area of risk management (such as mirror checks) back to the pupil when they're ready to take it on.

A score of 2 is good.

6. Were directions and instructions given to the pupil clear and given in good time?

This is a relatively straight-forward one to score well on.

Route directions need to be made clearly and in good time.

For example. "at the end of the road turn left", or "take the next turning on the right", or "pull up on the left at a safe place".

Instructions can mean "set the gas to 2,000" or "reduce your speed gradually to 20mph".

Avoid:

- potentially confusing or unclear instructions such as "slow down" or "turn right"
- giving directions too late.

Verdict: This is a relatively easy section to get right, so aim for a score of 3 here.

7. Was the trainer aware of the surroundings and the pupil's actions?

This is a key part of your role as an instructor. You have to be aware of what's going on around you and how your pupil is reacting to what's going on.

For example, you may be practicing left turns with a beginner and you have a car parked on the left in front of you and a lorry coming towards you. In this situation, you firstly need to be aware of the potential hazard and then instruct your pupil on how to deal with the situation.

On my most recent standards check test, a car was parked on the left blocking a zebra crossing. I missed a great opportunity to talk to my pupil about the dangers of blocking a crossing. I was so focused on the junction ahead that I didn't think of mentioning it. My examiner fed this back to be at the end of the test and I kicked myself for missing it!

Verdict: Difficult to score a 3 here, as the examiner will inevitably see things which you may not have missed or have deemed not worthy of mentioning.

Be happy with a 2.

8. Was any verbal or physical intervention by the trainer timely and appropriate?

Another tough one to get absolutely right. There's a very fine balance between saying too much and not saying enough. Whilst you need to be pointing out potential hazards early, you also need to allow the pupil to deal with situations that should be within their level of competence.

For example, if you have a meeting situation developing, you might intervene by saying "there's not room for two cars to pass here, so slow down and show me your holdback position". This is verbal intervention.

Physical intervention usually means using the dual controls in a potentially dangerous situation such as approaching a T junction too fast. In such a situation, you need to get the pupil to clear the junction and then stop them on the left and explain why you felt it necessary to intervene using the dual controls.

One situation I can recall was when my pupil panicked at an uphill T junction, which I had anticipated happening, as it's a notoriously nasty junction. I jumped on the dual controls and told my pupil to come off the clutch, as I had it under my control on my dual controls. Then give me 2,000 revs. We manoeuvred the car around the junction (with me using the dual controls) and then stopped at a safe place on the left. I calmed the pupil down and discussed what had happened and what to do if it happened again.

Avoid:

- Intervening unnecessarily (this can undermine the pupil's confidence)
- Not intervening when a pupil clearly needs your help.

Verdict: a fine balance and a tough one to get right on every occasion.

Be happy with a 2.

9. Was sufficient feedback given to help the pupil understand any potential safety critical incidents?

A safety critical incident is one, which could have potentially resulted in damage to a vehicle or injury to a pedestrian. These would be marked as a serious fault on a driving test report and the candidate would fail their test as a result.

Sometimes you can deal with faults whilst the pupil continues to drive, but for more serious faults (such as a right corner cut) it's essential that you instruct your pupil to pull over on the left and then discuss the fault with them in more detail.

You need to demonstrate your core competencies here, namely:

- Identify the fault
- Analyse it (discuss why it was potentially serious)
- Tell them what remedial action is necessary.

In short, ***see it, say it, sort it.***

It's important to allow your pupil to analyse the incident themselves, rather than you just telling them.

Using the example of a right corner cut, a good question to ask your pupil once they had pulled over on the left is "if you were doing that right turn again, how would you have done it more safely?" This gives them an opportunity to reflect on what they'd done (fault identification), why it was potentially dangerous (fault analysis) and what they will do differently next time (remedial action). Your pupil will learn faster if you let them analyse their own faults rather than you just telling them.

Verdict: be happy with a 2.

Section 3 Teaching & learning strategies

This section is about the way in which you teach your pupils. Your teaching style needs to meet the needs of the pupil, ie it needs to be pupil-centred. Some pupils respond well to verbal instructions, whilst others may be more visual and learn better if you explain a situation using diagrams. Other pupils may learn more from using actual examples. You need to use the right technique at the right time and encourage your pupil to analyse their own performance in a particular situation.

There are eight scoring areas in this section, so you can pick up a lot of marks by understanding what the examiners are looking for.

10. Was the teaching style suited to the pupil's learning style and current ability?

There are 3 main levels of teaching, namely:

- Full instruction (for beginners)
- Prompted practice (asking your pupil questions)
- Independent driving.

A complete beginner will need to be told exactly what to do, step by step. But as soon as they have moved off from the side of the road two or three times to your satisfaction, then you can start introducing some questions into your teaching style. For example, you might say: "well done Sam, you've moved off twice now with my help, now let's see how you get on with just a bit of prompting from me. So what's the first thing we do?"

Or for a more advanced learner, they may be ready to move up from you prompting them to more independent driving, so you may say: "OK, you seem to have mastered T junctions, have a go at them by yourself from now on. What I'd like you to do is to give me a running commentary of what you're doing and I'll give you some feedback afterwards. Is that OK with you?"

Your teaching style needs to match your pupil's learning style and their current driving ability. Try and progress them from full instruction to prompted practice and then onto independent driving. But if they're struggling a bit then consider going back to the previous level for a while until they appear more confident.

Avoid:

- Teaching at the wrong level
- Over instructing.

Verdict: A 3 is achievable here, but be happy with a 2.

11. Was the pupil encouraged to analyse problems and take responsibility for their learning?

Another relatively straight-forward one to score well on.

To score well on this point, you need to give your pupil plenty of opportunity to analyse what they did at a particular junction or on a reversing manoeuvre, analyse what they may have done wrong and what they need to do in order to put it right.

A technique I use frequently is to ask the pupil to pull over on the left at a safe place and then have a conversation with them along the following lines…

"If you were doing that right hand turn again, what would you do differently?

"Oh, I'd take it a bit wider".

"Yes exactly, what you did there is what we call a right corner cut (*fault identification*). What are the dangers of doing that at a busy junction?" (*fault analysis)*

"Well, a car could have been approaching the junction and the driver might not have seen me until the last moment".

"Quite right, so why would it be safer to take a wider line?" (*remedial action*)

"So that I can see down the road better, and cars coming up the road can see me earlier".

"Awesome!"

The key to scoring well in this section is to avoid simply telling the pupil what they did wrong. Without patronising them or boring them to death with over-analysis, encourage your pupil to analyse their own faults and to come up with their own solutions. In other words, give them every opportunity to teach themselves with just a little bit of guidance and steering from you. This way, their learning curve will be a lot steeper.

Verdict: I try and use this technique all the time with my pupils. Aim for a 3, but be happy with a 2.

12. Were opportunities and examples used to clarify learning outcomes?

- Tell a pupil and they'll remember 1%
- Show a pupil and they'll remember 10%
- Involve a pupil (using real examples) and they'll remember 50%.

Examples of this can be from your pupil's driving, or from what other drivers are doing.

A typical example of this is the white van driver. They can be very aggressive drivers for some reason. I frequently see them force their way around cars parked on their side of the road making you stop, even though it was your priority. I'd say to my pupil: "white van man strikes again. What was dangerous about their driving just there?"

It may not be particularly relevant to your particular lesson theme, but it's too good an opportunity to miss out on.

Seek out similar opportunities, which are relevant to your lesson theme.

For example, if your lesson theme is junctions, look out for other cars cutting right hand corners or incorrectly positioned at a T-junction for turning left. If you look, you'll inevitably find them. But you've got to be switched on and be looking out for them.

Verdict: this is 3 territory. Aim for a maximum score.

13. Was the technical information given comprehensive, appropriate and accurate?

This means giving the correct advice in relation to the theme of your lesson.

Example 1: When talking about your road position, the correct technical information is that you should be positioned in the centre of your lane on a normal road, or one metre from the kerb on a wide road.

Example 2: Adequate clearance: you should allow a one metre gap when passing parked vehicles at 30 mph in case a door flies open. If you need to leave less than a metre clearance, then you need to decrease your speed accordingly.

Example 3: Your normal holdback position should be 1.5 car lengths back from the obstruction and a metre from the kerb. This allows you to be able look down the road past the obstruction. It also looks like you're not parked on the side of the road, so cars approaching you from behind shouldn't try and pass you.

Example 4: During a reversing manoeuvre you should be looking through the rear window in the direction of travel for at least half the time, as this gives you the best opportunity to take appropriate action quickly should a pedestrian walk out behind you.

The key here is to research you lesson theme thoroughly. If your lesson theme is emergency stops, for example, then research the stopping distances of a car travelling at 30mph, 40mph, 50mph and 70mph. Also, what effect do wet and icy roads have on stopping distances?

Verdict: So long as you do your research thoroughly, a 3 here is very achievable.

14. Was the pupil given appropriate and timely feedback during the session?

Feedback means both positive feedback and constructive criticism.

Your feedback needs to be timely, so that the pupil can relate it easily to what's just occurred. Ideally give feedback whilst on the move, but if it's really important then ask your pupil to stop on the left and then give them your feedback.

Avoid giving vague feedback. A good example of what I mean is when teaching the emergency stop. Don't just say: "you need to brake harder", as what does this mean to the pupil…a bit harder or ten times harder? Instead be much more specific and say: "nice try, but you need to brake about 50% harder, so that we stop in a much shorter distance".

I also use the technique of what is politely termed a "crap sandwich". In other words, find two things that they did well and one thing that was not good. So, for example say: "you reacted really quickly when you did that emergency stop, which was great. You need to press the brake pedal about twice as hard next time, because we took quite too long to stop. Well done for remembering to put the handbrake on".

Occasionally you might completely forget to give feedback. This sometimes happens when I'm teaching. One example I can remember is when my pupil stalled the car on a busy roundabout. My immediate priority was to get the pupil to restart the car and to clear the danger area as quickly as possible. This occurred in a busy town centre and there was nowhere to stop on the left. By the time we were in a quieter area I had completely forgotten about the incident. I remembered later and we discussed it then.

If this happens on your part 3 test then it's never too late to recap what happened, even right at the end of your test. It's not ideal, but it's much better than simply ignoring it altogether. This actually happened to a fellow instructor colleague of mine on his part 3 test. He mentioned the fault right at the end of his test, about 20 minutes after the fault had occurred. He told me that the examiner had told him that if he hadn't mentioned the fault at all then he would have failed his test. As it was, he passed. It just goes to show that late is better than not at all.

Avoid:

- Over-criticising for the smallest mistakes
- Over-praising (praise then becomes less meaningful)
- Giving feedback long after the incident occurred (but this is still better than not giving any feedback at all).

Verdict: Aim for a 3, but settle for a 2.

15. Were the pupils' queries followed up and answered?

To score well, you need to prepare well and know the route you will be taking inside out. Careful planning can pay huge dividends here. When driving the route beforehand with your pupil, you can point out certain questions that the pupil may like to ask on the day of your test.

For example, if your pupil is approaching a zebra crossing, you may get them to ask you what the white zig-zag lines mean and then you give them a full technical answer.

Or..."can I overtake on the approach to a zebra crossing?"

"You can overtake a non-motorised vehicle such as a cyclist on the approach to a zebra crossing if you feel you need to, so long as it's safe to do so. But you must not overtake a car or any other motorised vehicle".

A word of warning here though: if you give incorrect or insufficient information this could lead to you failing your test. If you're asked a critical question and you don't know the answer (for example what a particular road sign means) then you need to say "I'm sorry, you've got me there. I don't know the answer to that question. I will find out and come back you this evening".

Ideally, give a comprehensive and correct answer. Failing that, say that you'll find out and get back to them. Never risk blagging it if you're not sure.

Verdict: Aim for a 3 here.

16. Did the trainer maintain an appropriate non-discriminatory manner throughout the session?

This is the easiest section to score a maximum 3 in, in my opinion.

What this means is…

- Not encroaching into the pupil's personal space (visualise a glass wall running down the centre of the car)
- Not saying any discriminatory remarks towards your pupil, pedestrians or other road users
- Not touching the pupil in any way
- Not waving a pencil of pen in front of the pupil
- Not making any derogatory gestures towards other road users (even drivers of white vans!).

You need to display complete impartiality towards everyone regardless of their age, skin colour, physical ability or disability and sexual orientation.

Don't even attempt to shake hands or do a high-5 with your pupil. I know it sounds daft, but just don't go there.

Verdict: an easy 3 points. A penalty in rugby union right in front of the posts!

17. At the end of the session, was the pupil encouraged to reflect on their own performance?

Another easy 3 points leaping out at you, so long as you remember to leave sufficient time at the end of the lesson for the pupil to reflect on their own performance during the lesson.

Try to use words of encouragement and avoid giving any destructive criticism. If your pupil did something particularly badly then turn this into an area that we need to work on next time. Ask open questions that encourage your pupil to speak openly, such as:

- "How do you feel that went?"
- "Which bits of today's driving were you most pleased about?"
- "Which bits shall we work on again next time?"

Then put these ideas into a framework for the next lesson.

Verdict: This should be another 3 points in the bag.

Preparing for your test

Find out which lesson themes work well at the test station where you're going to be doing your test. The best way to do this is to ask other instructors what worked well on their test and what lessons themes they would recommend.

For example, junctions at partly trained or trained level works well in Eastbourne, whilst a lesson based on roundabouts works really well in Hastings. Roundabouts is always the lesson theme I use for my own standards checks in Hastings. I have developed my own route, which I can adapt if necessary. This has worked really well for me in the past and scores well. Please contact me if you live in the South East and would like to know more.

Having decided on your lesson theme, choose a suitable pupil and adapt the lesson plan to meet their own particular needs. Make it challenging, but not overwhelming. For example, if your lesson theme is roundabouts, don't take a beginner on a multi-lane spiral roundabout! If you're on a PDI licence, practice the route over and over with different pupils. I would suggest at least 20 times. Learn the route inside out and what hazards may develop. For example, which sections of the route may be busy at various times of the day (eg school runs), or where the pedestrian crossings are, and so on. Know what to look out for on every part of the route.

If you're taking your test at the end of your ADI part 3 training then practice the route over and over with your instructor. Prepare thoroughly as if your life depends on it!

Pupils that fail are usually those who go out and try to wing it on the day. Practice makes perfect.

There's a big difference in the type of training your instructor should be giving you if you intend to take out a PDI licence compared to if you intend to take your test at the end of the 40 hours of ADI part 3 training. If you're planning on taking a PDI licence then your part 3 training should be predominantly on how to teach. You get a further 20 hours of training on your PDI licence, which can be used to prepare for your test.

If you're taking your test at the end of the 40 hours, then your training needs to incorporate much more preparation time for your part 3 test.

One size does not fit all. Your part 3 training needs to be adapted to meet your needs.

What happens on the day of your test?

I recommend that you park up close to the test centre with your pupil 20 minutes before your test is due to start. Have a final run-through with your pupil. Go into the test centre with your pupil 5 to 7 minutes before your test and wait in the waiting room. You will both need to have your driving licences with you.

The examiner will come and collect you. Introduce your pupil to your examiner in a casual way. You want your pupil to be at ease and as relaxed as possible.

The examiner will then ask you what your lesson theme is and will ask you about your pupil. The examiner will then have a chat with your pupil and ask him / her about their driving ability and how many lessons they've had. A good examiner will try and put the pupil at ease by thanking them for taking part in the test and saying that their driving is not being assessed in any way.

If your pupil is a beginner, then you may need to drive the car to a suitable area and your pupil will sit in the passenger's seat. Otherwise your pupil will drive, you will sit in the passenger's seat and instruct the lesson and the examiner will sit in the back and take notes.

Sometimes the examiner will be accompanied by another more senior examiner and this person will also sit in the back. Warn your pupil in advance that this may happen.

At the end of the test the examiner will go back into the test centre and write up their test report. They will come back to you after 5 to 10 minutes (this seems like an eternity) and go through their report with you.

KEY TIP: If you haven't been successful, then get as much advice and assistance from the examiner as you possibly can as to why they scored you lower than expected and what they expected to see from you. Ask the examiner about your choice lesson theme, your route, your teaching style. Get as much information from them as possible. This is absolute gold dust! Soak it up.

I have known an examiner spend as long as 20 minutes with one of my pupils in the car after the test, telling them what they needed to do next time in order to pass. But you've got to show them that you're hungry for their help and keen to learn from them, otherwise they may just give you brief feedback. So keep asking sensible and relevant questions. Then use this information to prepare thoroughly for your next test if you haven't been successful.

CHAPTER 5 – The 13 Biggest Mistakes to Avoid

In this section I will cover the 13 biggest mistakes instructor pupils make and how you can avoid making them.

Some of these mistakes I made myself and have proved very costly. Hopefully you won't make the same mistakes.

Mistake 1

Choosing your training company based purely on who appears to be the cheapest

There's usually a reason why the cost of instructor training is discounted: it's often linked to a 12 month or 24 month contractual tie-in.

Beware. Make sure you're comparing apples with apples when comparing prices.

For example, which is the better deal?:

Deal 1	Training only £995 Franchise fee £245 per week for a minimum initial term of 52 weeks
Deal 2	Training £1,995 No contractual tie-in.

Deal 1 will commit you to paying £995 + (£245 x 52) = £13,735 in your first year

Deal 2 will commit you to paying £1,995.

Let's look at the two options in more detail:

Deal 1 total costs are equivalent to you having to give 549 hours of paid instruction (at £25 per hour) before you start earning any money yourself! That's a huge financial burden to start your driving career with. Add to this your fuel costs means you've got to work long hours before you start to earn a living yourself. And what if you decide that being a driving instructor isn't for you? You may still be liable for making the monthly payments for the remainder of the contractual period.

Deal 2 costs are higher initially, but you're not committing yourself to a contractual tie-in. This is the route I went down after I qualified and I'm very glad I did. My up-front training costs were just over £2,700 and I lease my car, which I use for both work and for personal use. I simply put a roof box on top of my car when I'm giving paid tuition and I keep it in the boot when I'm not (it's magnetic).

My message is this: look at your total costs in the first 12 months when evaluating which training option to go with. Don't make your choice of training organisation based solely on the up-front costs.

Mistake 2

Believing you'll have a secure job at the end of your training.

You won't have a job. You'll be self-employed. Yes, it's a secure position, but it's NOT a job.

Many instructors I know had left corporate life for one reason or another to become driving instructors and were under the impression that once they'd completed their training they would have the comfort of a secure job with a guaranteed income. That's the life they had been accustomed to all their working lives and they liked the idea that their diary would be managed for them and that all they had to do was to turn up for pre-set appointments.

It's most definitely not a job. You're on your own. You're self-employed.

In my opinion being self-employed is preferable to being employed. I'd run my own businesses before, so it was nothing new to me. I love the freedom and control that being self-employed offers and I doubt I would ever go back into paid employment again.

Mistake 3

Believing the ads that state you'll earn £30,000 to £50,000 per annum

Ads like this make my blood boil (taken from the first page of Google search results):

1. Become A Driving Instructor | Earn £30k In Your 1st Year

Adwww.xxxxxx.co.uk/First_Year/Earn_£30k

Guaranteed £30,000 Income. As A **Driving Instructor** Or Your Money Back. Apply Now

That's like telling a greengrocer they will earn £30,000, but forgetting to deduct the wholesale costs of the fruit and veg!

It's the **NET** amount that's really important, not the gross income from lessons.

Your **NET** earnings are your income from driving lessons **MINUS YOUR COSTS.**

Then deduct tax and national insurance.

If you're looking to make £30,000 net income by taking out a franchise, then your GROSS income needs to be £30,000 **PLUS YOUR COSTS**.

So, using a franchise fee of £245 per week, this comes to £30,000 + (£245 × 52) = £42,740. Add to this your fuel costs of £3,500 = £46,240.

At £25 per hour, this means that you will need to work (£46,240 / 25) hours per annum

= 1,849 hours per annum

= 40.2 hours per week (based on 46 working weeks per annum).

At first glance that doesn't seem too bad. But this assumes that you live in a perfect world where there's no bad weather, no lesson cancellations and all your lessons are back-to-back.

Doing the same calculations to earn a net income of £50,000 a year before tax means that you'll have to work (£50,000 + (£245 × 52) + £3,500) / 46

= £66,240 / 25 hours = 2,649 hours per annum

= 57.6 hours per week (based on 46 working weeks per annum).

I don't know any driving instructor who consistently works that many hours per week over a sustained period of time. That's equivalent to around 12 hours per day, 6 days per week, every week, summer and winter.

To make it worse, the ad even says it's a **GUARANTEED** income of £30,000 to £50,000!

How much you earn depends on how many hours you work MINUS your overheads. There's no holiday pay. You don't get paid if the weather's bad or if you're ill or if your car's off the road. There's no pension provision (that's completely down to you) and you most certainly don't get paid overtime for working in the evening and at weekends.

Mistake 4

Believing qualifying as a driving instructor is straight-forward

It's not. It's a lot more difficult than people imagine.

The overall pass rate for ADI Part 3 exams in 2017 was a mere 27%.

That's shocking enough. But it gets worse:

The pass rate was even lower than this for the first attempt at the part 3 test

One of the major concerns highlighted by the DVSA is why so many people who set out to become driving instructors never actually qualify. Some will simply decide that it's not for them during their training process (after having parted with considerable training costs). Others will fail to qualify, usually at the final hurdle of their part 3 test.

Qualifying is NOT straight-forward. Of course it can be done, but it requires a lot of effort, determination, high quality training and a lot of practice.

In my own training, I changed my training organisation and trainer three times before finally passing my part 3 exam. Had I not taken such action I would not have qualified. Sadly, this happens all too frequently and many aspiring instructors don't reach the finish line.

Mistake 5

Paying for all your training up-front in a single lump-sum payment.

This is risky, unless you know exactly what you're going to get.

- How do you know for sure that this is the right career for you?
- Does the quality of the training provided meet your expectations? (mine certainly didn't)
- What is the support like during training? (mine was virtually non-existent)
- Is there a specific trainer allocated to you? If so, what is his/her availability?
- If your circumstances change and you have to pull out, do you get a refund?
- If you don't pass your ADI part 1 or part 2 tests do you get a refund?

So what's the alternative?

I've thought long and hard about what I think is the best payment structure that's fair for both the training company and for the trainee. For my company's training, 25% of the total training cost is paid to reserve your place on the course. This covers the cost of the training and support up to and including the ADI part 2 test. After this point, my pupils are free to decide whether they want to continue or not.

This way, my pupils are not risking losing a significant amount of money before they have gained a good understanding what being a driving instructor is all about and whether it's right for them. Those that decide to continue and train for their part 3 test (the vast majority of my pupils do) then pay the balance of their training costs.

Mistake 6

Taking your first ADI part 3 test before you're 100% ready.

That's what so many trainees do (myself included). Because I was none the wiser. My original training company didn't even mention the PDI option, so I simply went ahead with my first attempt at my part 3 test and I completely wasted it. I was like a rabbit caught in headlights. I wasn't ready and I'd had no experience with real pupils. I just went ahead with what I was told I had to do.

The big mistake is thinking you've got three attempts at your part 3 test, so you might as well have a go and see how you get on.

NO!! Don't do that. And don't let anyone tell you that's what you've got to do.

Those three attempts are precious and you don't want to end up on death row like I did with just one attempt remaining. That was not a pleasant experience and I really felt the pressure. Fortunately, my third and final attempt went like a dream and I passed. But that was primarily because I changed my trainer, which is another story.

If you've completed your 40 hours of part 3 training but you don't think you're ready to take your part 3 test then I suggest you do one of three things:

Either

- pay for a further block of lessons with your existing trainer, or
- pay for an additional block of lessons with a different trainer, or
- take out a PDI licence, which gives you up to 6 months experience of teaching pupils.

In my own case, I took my part 3 test straight after I had completed my 40 hours of training and I failed it miserably. I then changed my training organisation, took out a PDI licence for 6 months and bought two blocks of further lessons at considerable extra cost. Yes it was extra cost, but had I not taken the action I did then I am convinced I would not have qualified. I found the experience of teaching real pupils and receiving the highest quality training invaluable.

Mistake 7

Having pay-as-you-go pupils

I've heard every excuse under the sun as to why pupils can't make their lesson today. Amazingly, many of my former pupils used to get colds on really hot sunny days. I'm sure it had nothing to do with them perhaps just wanting to go to the beach!

I now run driving courses, not driving lessons. Just a little tweak, but it makes a big difference.

My pupils pay up-front for a block of at least 10 hours of lessons. I give them the terms and conditions regarding my cancellation policy in black and white on their first lesson, so that there is no argument. My pupils know exactly where they stand. I also text my pupils the evening before their lesson with a reminder. This takes no more than two minutes and it works well.

It's funny how quickly some pupils recover from being on their death bed when they realise they'll lose their money if they cancel their lesson within 24 hours!

Be firm but fair with them. Don't stand for any nonsense. If a pupil has been with me for a long time and I know that their reason for cancelling is genuine and this is a very rare event then I would take a view and probably not charge them on the first occasion this happened. But I would make it clear that this is the first and last time I'll not charge them for a last-minute cancellation. Most pupils completely understand.

I get very few last-minute cancellations now. If a pupil does cancel at the last minute then I rarely lose out financially as the lesson has already been paid for. I get zero no-shows.

My message is this: get pupils to pay in advance. Ideally for a block of 10 hours of lessons or at the very least charge them for their next lesson in advance. Say that it doesn't go in the diary until it's paid for.

If a pupil is not prepared to pay in advance for their lessons then think seriously about whether you want to take them on or not. I choose not to. All my pupils pay in advance for a block of lessons. Usually for 10 hours or more, but occasionally for 6 hours. This way I filter out any pupils who are likely to prove unreliable in the future and as a result my last-minute cancellation rate is extremely low.

Mistake 8

Relying solely on one single source of leads

Don't put all your eggs in one basket and rely on just one source for new pupils.

If you're relying on a call centre, for example, what would happen to your business if the call centre closed down? By all means use it, but don't rely solely on it.

Assuming you're paying for this service, ask yourself why you need to continue to do so in the future if there were other ways of getting pupils for little or no cost.

Take responsibility for your own marketing and try three or four of the marketing ideas outlines in the chapter on Marketing Essentials.

There are a lot of different ways that you can market your business at little or no cost. If these don't prove productive then try other ways. Be prepared to experiment and evaluate.

But don't rely solely on just one method for finding pupils.

Mistake 9

Offering up-front introductory offers.

I quite often see other driving companies offering up-front introductory offers such as:

"First 5 lessons for £60", or "First 2 lessons for £10".

These can attract the wrong type of pupils, who are just jumping from one introductory deal to another. Such discounts are usually unnecessary anyway, unless you're working in an area where competition for pupils is high.

When quoting my price over the phone, I always quote for a block of 10 hours of lessons. I say something like:

"My standard rate is £30 per hour, but most people buy a block of 10 hours for £250, as it saves them money".

That's the type of pupil I want. They tend to be much more reliable. Having paid in advance, they're much less likely to cancel at the last minute. And if they do, the lesson has already been paid for.

Some pupils asking for cheap deals will take you up on your offer and then you don't hear from them again, as they've moved on to the next introductory offer.

If you want to offer an incentive, then make it a back-end incentive.

For example: "buy a block of 20 hours and use my car during your test for free".

Look at what I'm offering: buy 20 hours of lessons from me and at the end I'll give you 45 minutes use of my car for free. The free bit is at the end, not up-front.

That's a much better way of offering an incentive. You get a long-term commitment from them first and then you give them something back at the end. Just like you would leave a tip in a restaurant AFTER you've had your meal.

Mistake 10

Being at the beck and call of your pupils 24/7

Be firm but fair with your pupils. Let them know where you stand from the outset.

I take calls from pupils in the evenings or at weekends only if it suits me. Sometimes it does and sometimes it doesn't. But I make that decision. My pupils know that and they accept that I do have a life away from driving as well.

You'd be surprised when people call you. I once received a call from a lady enquiring about driving lessons for her daughter at 23:55pm on a Saturday night! Her call went through to voicemail and I returned her call on Monday morning. To be fair, the lady wasn't expecting me to answer her call when she rang. She was expecting it to go to voicemail.

I'm sure that some pupils live in their own world where they think they're the only pupil I teach and that I'm available 24/7. It's a case of managing their expectations and finding a way that works for both you and the pupil.

I encourage my pupils to text me, because I can choose a convenient time to read their text and answer it. This works well. I don't mind being texted in the evenings and at weekends and I do try and get back to them promptly.

Similarly with regard to diary clashes. This is a typical example:

Your pupil rings you up and says "I've just booked my test and it's on 3rd September at 8:30am".

It's your daughter's first day at her new school. It's been in your diary since June.

What do you do?

I would tell them I'm not available. My daughter would come first. They'll have to reschedule their test. And they do.

Mistake 11

Focusing solely on your pupils as a PDI

This mistake is so easy to fall into if you go down the route of taking out a PDI licence before taking your ADI part 3 test.

The main purpose of taking out a PDI licence is to give you experience of teaching real pupils before taking your ADI part 3 test. But this can actually hinder your chances of passing, as opposed to helping you. The reason being that you can concentrate so much on running your business and teaching your pupils to pass their tests that you forget the main reason for taking out the PDI licence in the first place.

This is a mistake that I constantly remind my instructor pupils about. It's like being on board an aircraft and you have a small child with you and you have to fit oxygen masks. You are told in the safety briefing to fit your own mask first and THEN fit the child's mask on. Hence, look after yourself first and then help others.

It's exactly the same on a PDI licence. Practice your part 3 route over and over with your pupils. Get to know the route over and over. Get to know every turn and where your pupil is most likely to make a driving fault when you're doing your ADI part 3 test.

I can relate directly to this from my own experiences as a PDI trainee. I became good at getting pupils through their driving tests and my monthly accounts were always bang up to date, but I hadn't practiced anywhere near enough for my ADI part 3 test using my pupils. Before I knew it, the date of my second ADI part 3 test was upon me and I blew it. I then changed my trainer again to someone who knew the system inside out. He badgered me to give him a written report every week detailing what ADI preparation work I had done every day with my pupils and in my own time. I practiced for hours and hours every day with my pupils and it worked big time. I passed my part 3 test on my third attempt with just one mark short of a perfect score.

I apply exactly the same technique when helping my own pupils prepare for their part 3 test. During their instructor training, it becomes apparent what their strengths and weaknesses are. Some instructor pupils are very good teaching new driver pupils clutch control, moving off and stopping. Others are more natural teaching pupils who are closer to test standard. Others may be particularly strong teaching reversing manoeuvres. We focus on their strengths and develop a lesson plan for their part 3 test, which plays to these strengths. Then we find a suitable pupil and adapt the lesson to meet the particular needs of that pupil. We practice it over and over with different pupils. That's why our pass rate is well above average. Practice makes perfect. You can't wing it on your part 3 test. You have to be fully prepared.

Exactly the same principal applies to qualified ADI instructors when preparing for their standards check test. The theme of my lesson on a standards check test is always roundabouts. I have a fixed route in Hastings, which I know takes 44 minutes to complete on a clear run and 48 minutes if traffic is heavier than normal. I've had over 50 of my pupils drive round the route as part of their normal driving lessons, so I know the route inside out and how to shorten it if the need arises. I also know through experience what other driving faults to look out for and how to adapt the lesson to deal with these faults should the need arise.

To quote the legendary golfer Gary Player, considered to be the master of bunker shots, "it's funny you know, the more I practice, the luckier I get". You should apply the same principal in preparation for your part 3 test: "the more pupils I practice my part 3 exam test on, the luckier I get".

The message is this: your number 1 priority as a PDI trainee is to get as much relevant experience as you can to help you pass your ADI part 3 exam.

Mistake 12

Doing the wrong type of ADI part 3 training

Here I am at odds with some of the main driving instructor training schools, who will disagree with what I'm saying here...

In my opinion what you focus on in your 40 hours of part 3 training should depend on whether you're going down the PDI route or taking your ADI part 3 test at the end of the training. The two routes are completely different.

Many of the large training schools say that the training should be the same irrespective of which route you decide to go down. One instructor training company I teach pupils for state that the first 4 days of the intensive training course should be used for teaching instructors how to teach their pupils and only the last day should be used for test preparation. I disagree entirely with this. It's complete madness in my opinion.

If an instructor pupil is going down the PDI route then the prime focus of their training should be on how to teach pupils. They get a further 20 hours of training in their first 3 months as a PDI and these 20 hours are ideal for preparing them for their test. Why prepare them for their part 3 test any earlier if they're not taking their test for at least another 4 months?

On the other hand, if you're planning on taking your part 3 test at the end of the 40 hours of part 3 training then a much greater proportion of the 40 hours should be allocated to preparing you for your test. The prime focus of your training is learning to pass your part 3 test at the end of it. What's the point of using up hours of your precious part 3 training time learning how to teach reversing manoeuvres if you're not going to use this skill on your part 3 test? Sure, I cover reversing manoeuvres briefly, but not in any great depth.

In my opinion this is a major reason why the ADI part 3 pass rate is so appallingly low (a mere 27% pass rate in 2017)... many instructors simply delivering standard part 3 training instead of delivering bespoke training based on the needs of each individual instructor pupil. Your part 3 training must be tailored to your own particular needs. One size does not fit all. This is why the pass rate of our own pupils is significantly higher than the national average.

And last but not least...

Mistake 13

Paying for the privilege of advertising someone else's business

Why do you want to pay good money out each week to advertise someone else's business all over your car?

You want to advertise **YOUR** business, not someone else's. Maybe not right away from the word go, but that's where you should be aiming for after say 12 months.

And it's your car. You're paying for it, so why not make full use of it for your own personal use as well as for giving driving lessons? Do you really want to take your family and friends out in a car with the name of a national driving school splashed all over it in big bright red and yellow letters?

I just have a magnetic top-box, which I put on top of the car when I'm teaching pupils. When I'm not, the top-box lives in the boot of my car and it looks like a normal everyday car.

Aim to create your own brand at the earliest opportunity and certainly after a year. Build your own business. Get a constant stream of referrals coming in from happy customers.

Why do you think over 90% of all instructors who have been in the industry for more than four years have their own driving business? You see it on their car top-boxes...

> "Alison's Driving School 07775 404412"

> "East Sussex Learners 01424 888333"

The answer is because they have realised that there's plenty of demand for driving lessons and that they can manage by themselves and thereby cut their overheads to a minimum. They're building their own businesses and getting new enquiries coming directly to them. Happy days!

CHAPTER 6 –
How to Start Earning Whilst You're Training

In this chapter, I'll explain the two different paths that you can go down after you've passed your ADI part 2 test and I'll go through the advantages and disadvantages of each option.

The two options open to you after you've passed your ADI part 2 test are:

Option1

Complete your 40 hours of part 3 training and take your ADI part 3 test at the end of the 40 hours.

Option 2

Complete your 40 hours of training and apply for a PDI licence (trainee instructor's licence). The licence costs £140 and lasts for 6 months, during which time you can not only practice for your ADI part 3 exam with real pupils, but also charge full price for giving lessons. So this way the can start earning a living whilst you're still training.

Some training organisations offer both options. Others don't. I recommend you keep both options open and make an informed decision, rather than only having one option available to you from the outset. Make sure your training provider offers both options. If they don't, then consider finding an alternative training organisation.

So what are the pros and cons of each option?

Option 1 : take your ADI part 3 exam at the end of your training

The main benefit of this route is that you can qualify as an ADI at the earliest opportunity. Many pupils want this absolute certainty before resigning from their current job. Remember that there's no guarantee that you'll pass your ADI part 3 exams, so resigning from a salaried job before you've qualified can be risky in certain circumstances. You also don't incur the cost of a PDI licence (currently £140).

About a third of the pupils I have trained have gone straight into their part 3 test at the end of their 40 hours of training.

Option 2 : complete your 40 hours of training and apply for a trainee instructor licence (PDI)

The two main benefits of this option are firstly that you can start earning money much sooner than if you were waiting to take your ADI part 3 test. The second huge benefit is that this option gives you valuable experience teaching real pupils, and you can use your lessons to practice for your part 3 test.

There's a lot of information about the rules regarding the PDI licence on the .gov.uk website and rather than regurgitate it all, please have a look at the following link...

Get a trainee driving instructor licence - GOV.UK

https://www.gov.uk/trainee-driving-instructor-licence-the-rules

You can **apply** for a trainee **licence** if you: have passed your ADI part 1 test in the last 2 years. have passed the ADI part 2 test. have had at least 40 hours of training from a qualified ADI in providing driving instruction (at least 10 of which were done in a car), recorded on the ADI 21T declaration form.

Apply for a trainee driving ... · Options when you apply for a ...

This option suited me, as I wanted to start earning as quickly as possible and also maximise the likelihood of passing my ADI part 3 exam. About two thirds of the pupils I teach take the PDI licence route. Some of them would have struggled to pass their part 3 exams without having the extra training and experience that this option offers.

In my opinion, the main reason why the pass rate for the part 3 test is so low is because too many pupils are forced to take their part 3 exam at the end of their 40 hours of training when they're clearly not ready, because their training provider didn't offer them the PDI option.

But there are drawbacks to the PDI option...

Firstly, you may have to give up a salaried job to become a PDI, but you could be out of work within 6 months if you don't pass your part 3 test during this period. In other words you're making a big commitment before you've fully qualified.

Secondly, what do you do about a car for giving driving instruction? I took a chance and leased a Vauxhall Corsa on a 12 month contract, even though I was only certain of an income for the first 6 months. I have since discovered that CA Cars in Leicester offer cars for driving instruction on a 6 months lease, which is ideal for anyone going on a PDI licence. I suggest you get in touch with CA Cars and find out what deals they have available if you're considering going down the PDI route.

Thirdly, there is the additional cost (currently £140) of a PDI licence.

So each to their own. Horses for courses. The important thing is that you know that the PDI option is available. It's essential that you're at least made aware of its existence from the outset.

Ask your prospective training organisation if they offer this option. If they don't, then consider finding an alternative training provider. It's risky putting all your eggs in one basket and not having the PDI option available to you. Instead, try to keep both options open for as long as possible and then make an informed choice. With my own pupils, in nearly every case, they are clear which option is best for them by the middle of their ADI part 3 training.

CHAPTER 7 – Fast-Track Training

In this section, I will explain how some of my pupils have started earning an income as a driving instructor within four months from starting their driving instructor training.

It's a fast-track process, which requires a great deal of flexibility from your training provider.

This is a relatively short chapter, but it's a really important one if you need to start earning money as a driving instructor quickly.

Many people looking to become driving instructors need to start earning money as soon as possible. Maybe they've unfortunately lost their job, been made redundant or taken early retirement. They've still got bills to pay and mouths to feed.

The industry standard for completing the training is between 9 months and 18 months. For many pupils this is far too long. Most instructor-training companies have set procedures and timescales for their training courses. They're not particularly flexible to each pupil's needs.

From a financial point of view, if you have no other income stream whilst you're training, you're effectively MISSING OUT on around £750 of gross income from driving lessons per week (based on 30 hours of income at £25 per hour) for every week that your training is taking.

So, if your training takes 4 months before you start earning, then you're limiting what you're missing out on to (17.3 weeks x £750) = £13,000 gross income (before deducting your costs).

If your training takes 9 months, then this figure increases to £29,250 gross income.

If your training takes 18 months, then the figure rises to a whopping £58,500 in potential lost gross earnings.

Your training does not need to take between 9 and 18 months. There is a quicker way. Here's how some of my pupils have **started earning within 4 months** of first contacting me...

The 10 steps in my fast-track training process are as follows:

Step 1	Complete a DBS check
Step 2	Get a PRN (personal reference number) from the DVSA
Step 3	Book and pass your ADI part 1 (theory test) at your local test test centre
Step 4	Book your ADI part 2 (practical driving test)
Step 5	Complete 10 hours of training for your ADI part 2 test
Step 6	Pass your ADI part 2 driving test
Step 7	Complete 40 hours of training for your ADI part 3 test
Step 8	Obtain a PDI licence
Step 9	Join a local driving school close to you (we help you find one)
Step 10	Start earning.

The four key steps to achieving this fast-track process are...

Key Step 1: **book your ADI part 1 and 2 tests at the earliest opportunity**.

Don't wait until you've completed your training. Get on the .gov.uk website, book your test and THEN do the relevant training.

Key Step 2: **do as may of the above activities in parallel**.

For example, some my pupils on the fast-track programme do their part 2 training at the same time as they're training for their part 1 theory test.

One of my pupils did exactly this. He passed his part 1 theory test and then went straight onto the .gov.uk website and was fortunate to find a part 2 test cancellation date within a week of passing his part 1. As he had already completed his part 2 training, he booked his part 2 test and passed it a week later with just 4 minors. He was fortunate to find a test date so quickly, but it's proof that it can be done.

Key Step 3: do an intensive course for your part 3 training.

Many of my pupils travel from all parts of the UK and stay locally in an Airbnb or a Travel Lodge. We start the part 3 training at 8am on a Monday morning and finish at 5pm on a Friday afternoon. 40 hours of training completed in 5 days. It's mentally exhausting, but it can be done and I've done it with some of my pupils.

Key Step 4: obtain a trainee licence (PDI).

After you've completed your 40 hours of part 3 training, obtain a trainee licence (PDI badge). This normally takes around 2 weeks to come through.

You can then start earning an income and at the same time gain valuable experience with real live pupils. This is hugely beneficial to help you pass your ADI part 3 test.

This is what I did when I was training. I joined a local driving school (Time2Pass in Hastings) and they provided me 20 pupils within the first 2 weeks. My diary was full in no time. As well as getting paid to teach pupils how to drive, I was also preparing for my part 3 test at the same time.

This fast-track route isn't suitable for everyone, but I offer it as an option for those who need to start earning quickly.

For further information and to check availability please email me at bc@briancroucher.co.uk

CHAPTER 8 – Marketing Essentials: How to Get New Pupils for Little or no Cost

Marketing Essentials - Proven Ways to Attract New Pupils for Little or no Cost

It's really important that you generate your own leads, irrespective of any franchise deal that you've signed up to. Here are some proven ways to get new pupils for little or no cost. You don't need to adopt all of the methods below, just pick three or four and give them a go.

Don't rely on any single source of leads. Try and find at least three different ways that work for you. Experiment and see what works. If it's not working then try something different. Here are some proven methods that work...

1. Branded Polo Shirts

These work brilliantly in places such as supermarkets, where there are a lot of people in close proximity. Keep your message simple. For example, on the reverse of your shirt have the words "Driving Instructor" stitched on. Then on the front just have your name and the words 'Driving Instructor" underneath. Choose your own colour schemes, with contrasting brand stitching.

To find suppliers, do a Google search on "branded polo shirts".

I've used www.WorkwearExpress.com in the past and their service was fine. Go to their customisation section and take your pick on colours and designs.

You'll get stopped frequently in busy places such as supermarkets, so always keep a supply of business cards with you and ***ask them for their name and mobile number***.

You're ten times more likely to get a new pupil if you take the initiative and call them back rather than waiting for them to call you. You've got every reason to call them back, so I would suggest doing so the same day. Strike while the iron is still hot.

2. Facebook Business Page

In my experience, having a Facebook page is far better than having a website. The reason being that it's quick and simple to like a page and to spread the word. A website requires extensive search engine optimisation, whereas a Facebook business page doesn't. The majority of your pupils use Facebook every day, so if your Facebook page is set up correctly it can provide you with new leads in a short space of time.

The best things to post on your Facebook page are photos and videos of pupils immediately after they've passed their tests. I always take at least one photo of my pupil in front of my car (with top-box clearly visible in the background) clutching their test pass certificate. Most pupils will be happy to give you a quick 20-second testimonial video as well when they've passed, which you can upload to Facebook. Make sure you leave your contact details for their friends to contact you.

If you want help setting up a Facebook business, ask a teenager!

3. Intensive Courses

Become a local instructor for one or more of the many national companies who market intensive driving courses.

This is a really good way to get paid work once you're a fully qualified ADI. I use this method a lot. The beauty of this is that the pupils all want to learn quickly and pay up-front for their entire course before they start, so there's never any issue about getting paid.

To find out which companies advertise in your area (say Newport), do a Google search on "intensive driving course Newport". Write down the names of the ones who appear on the first page of the search results and click on their links. Look for a link such as 'Instructor Login' and register with them. You will need to email them a copy of your driving licence and ADI badge. They're all free to join.

Examples of such companies are:

- midrive.co.uk
- oneweekdrivingcourse.co.uk
- fastpassintensivecourses.co.uk
- passmefast.co.uk
- momentumdrivingschool.co.uk

4. Get some business cards printed

A no-brainer. You must get some business cards printed.

Do a Google search on "business cards" and take your pick.

I use vistaprint.co.uk but there are loads of companies to choose from.

Keep your message simple. Just include your name, your title of 'Driving Instructor', your mobile number and email address.

Keep the reverse of the card blank, so that people can write on the back. Don't go for glossy cards, as it's a pain trying to write in ballpoint pen on them. Just plain matt cards are fine.

5. Get some Leaflets printed for pupils to give away to their friends

Offer an incentive on the leaflet, such a £25 off when you book a course of 15 hours or more (in other words you're giving them one hour free for every 14 they pay for).

The leaflets don't need to be large. A6 size is fine. Just the right size to stick on a fridge door with a bit of Blu-Tac. Sometimes you don't hear back from their friends for several months, but I usually get at least one new lead from every pupil who passes their test with me. Either through Facebook or by them phoning the number you wrote on the leaflet. This keeps a regular supply of new pupils coming in for me.

6. Advertise in your local corner shop

The best way to do this is to buy some cheap L plates and write your name and mobile number in black marker pen in the white space on the L plate. Anyone interested is most likely to take a photo of it on their phone and call you.

Advertising this way typically costs 50p to £1 per week.

The good thing about advertising in your corner shop is that any leads that arise are likely to be local to you.

When they contact you, they will normally ask you about driving LESSONS. I explain that I provide driving COURSES, of which there are three to choose from:

- New driver course (for someone with little or no previous experience)
- Intermediate driver course (for someone who's driven a bit before)
- Test preparation (for someone approaching test standard).

Each course normally takes between 8 and 12 hours, depending on the pupil. Any unused hours can be carried forward into the next course.

That's the mindset that you want to get over to your pupils (or to the parents of the pupils if they are the ones paying for their son/daughter's course). It therefore comes as no surprise to them when you tell them about the prices for a block of lessons, as opposed to the prices for individual lessons.

7. Advertise in Busy Car Parks

Try and park your car in a prominent position in a busy car park (eg Tesco's), as close to the main doors as you can get. Leave your instructor's top-box on the car and leave some of your leaflets in a clear plastic box that clips to your door window (these cost around £5 from online retailers such as ebay and Vistaprint).

Supermarket car parks are excellent places for picking up new leads, as people only tend to stay for short periods of time, as opposed to all day in other car parks. So you get a higher throughput of potential pupils (or parents of potential pupils).

Gift vouchers also work well in the run up to Christmas and birthdays. You can create your own vouchers or get them printed. Be careful that they can't be photocopied or duplicated. I normally sign on the back of them or personalise them with the pupil's name before giving them out.

Sometimes I don't even print off the vouchers. The person buying the vouchers transfers the relevant amount of money into my bank account and I send them an email back confirming how many hours they've bought and how to contact me. Sometimes that's all that's needed.

8. Buy Leads

Let's say you live in Cambridge…

If you want to buy leads that have gone to other instructors, do a Google search for "driving lessons Cambridge" and see which local driving schools rank high on the first page of Google. Ignore the big nationals such as Red and AA, as they won't be interested in selling you any leads.

Talk to three or more local driving schools that appear on the first page of Google. Be upfront with them say that you're a local driving instructor looking for a regular source of new pupils. Ask them what they do with any leads that they can't follow up quickly themselves.

Offer to buy any unused leads from them for a 'fair price'. A fair price would be around £20 for each lead that converts into a sale.

Such an arrangement will rely on mutual trust, so be honest with them, in order to maintain a steady flow of new pupils coming to you. You don't really want to start building your own website until you've become established, especially if you're on a PDI licence.

This is a great way to get started and it's how I got started when I took out my PDI licence. I received an unlimited supply of leads, for which I paid £50 per week. I remain good friends with the owners of the driving school that helped me get started.

9. Do a lead swap

If you offer lessons in a manual car, for example, find out who offers lessons in an automatic and phone them up. Ask them if they would be interested in you sending them leads for automatic lessons in return for them sending you leads for pupils wanting to learn in a manual car. This is completely free of charge, as it's simply a lead swap.

10. Advertise in yell.com

I haven't personally used this method, but one of the lady instructors I taught in the past told me that she gets nearly all her enquiries through yell.com. It costs her about £75 per month. Worth checking out.

11. Have Breakfast at your local Federation of Small Businesses (FSB) meeting

Find out where your local FSB meeting is held and go along to one of their local business breakfasts. These take place once a month and they're usually well supported.

Don't try and sell your services to the room, instead sell through the room. In other words try and reach 17-20 year olds whom they may know.

Such networking events are a great way of getting to know local business people, for example web designers, social media trainers and printers. You get a chance to stand up and talk about your business for about a minute, so grab this opportunity.

Here's the pitch I use whenever I go. Feel free to copy it, as it really does grab people's attention while they're eating their breakfast.

Stand up and act very sombre. Speak slowly and say:

"Good morning. I need your help please. I keep losing my best customers. I lost 2 last week and 3 the week before. Please can you help me?".

Then pause and in a very energised state say:

"You see I'm a driving instructor and keep losing my pupils, because annoyingly they keep passing their flipping driving tests".

Pause to allow the chuckling to die down, then say:

"Who do you know who may be approaching their 17th birthday and looking for driving lessons? I've got some money-off vouchers here and I would I be very grateful if you would take a few and hand them out to people who know who might be interested in passing their test. Hopefully at their twelfth attempt. Thank you"

Trust me, that pitch always goes down well. Practice it beforehand. It's OK to just stand up and read it out if you don't particularly enjoy public speaking.

12. Have your own website built

Last but not least, you will need your own website if you intend to build a substantial business and take on other driving instructors. However this is NOT the best use of your time or money when you're just getting started. A decent website will cost you upwards of £500. Then you'll need regular SEO and maintenance work to keep it up the Google rankings.

I'm not a great fan of using websites. In my opinion, using Facebook is a much cheaper and better alternative.

Finally, a little rant...

If you're going to start your own driving school, then good for you. Think very carefully about your potential customers, namely 17 to 20 year olds.

Come up with a name that will appeal to them. But please DON'T call your company "XXXX School of Motoring".

Why not?

Because 'Motoring' is a very old-fashioned word and conjures up an image of a 65 year old instructor smoking a pipe.

Now the word "School": What do most 17 to 20 year-olds detest and have just recently left?

School.

So don't call your business a 'school'.

How about 'Driver Training'? The name portrays them behind the wheel of a car learning to drive. It's a cool thing for them to be doing when they reach 17.

Rant over. Hope it makes sense.

CHAPTER 9 – The 7 Keys to SUCCESS

In this section I will explain what I believe are the 7 keys to success, both for qualifying as an ADI and for running your own driving instructor business.

1. Know your 'Why'.

Why do you want to become a driving instructor?

Take a few minutes to write down at least 5 reasons why you want to become a driving instructor and list the benefits of each one for you and your family. The more reasons you come up with, the bigger your 'why'. The bigger your 'why' is, the more likely you are to succeed.

Try and think of the benefits from a positive perspective. For example, don't say 'I hate having to commute at silly o'clock every day and getting stuck in traffic'. Instead, a more positive approach would be 'I am so looking forward to having quality time with my family first thing in the morning'.

Some of the compelling reasons for me when I first started were...

- Complete control of my time. Work when I choose to, giving me complete flexibility and freedom
- Ability to take time out to go to Australia every year and spend quality time with my children and grandchildren who live there (this is a massive 'why' for me)
- Be my own boss, with absolutely no fear of losing my job
- Do I job I enjoy
- Stress-free work.
- Start work within 15 minutes of leaving my home. No long commute.
- An opportunity to meet new young people and make a difference to their lives
- A good reliable income
- Play golf on Friday afternoons.

Take time out now in a quiet place, away from the TV and write down your own reasons why. Keep going for at least 30 minutes. Just write down exactly what comes into your head. Chances are you'll be amazed at what comes out of your mind the deeper you go.

2. Do your research thoroughly

Do a Google search "become a driving instructor". This search will list at the top the companies who operate nationally. Then click on the links on the first page.

For example...

www.theaa.com

www.reddrivingschool.com

www.billplant.co.uk

www.smartdriving.co.uk

Request information packs or enter your details on their 'Contact Us' page.

Then do a second Google search for your own area. For example if you live in Hull, do a Google search "become a driving instructor in Hull". This time look for the companies that are specific to Hull.

So you should have 2 lists:

- List 1 are the national companies, such as the AA
- List 2 are the companies who operate locally in your area.

For each of these training companies find out and compare:

- The cost of their training
- Where the training takes place (do they have a trainer near you?)
- Do they have a contractual lock-in period at the end of the training?

(If so, how long is it for and what is the total cost per month?)

- What's included in the cost, and more importantly what isn't?
- Do you have to pay for the entire training course in one lump sum?
- What guarantees or money-backs do they offer, should you not want to continue at any point or not pass your tests?

Another great place to do your research is on YouTube (YouTube is actually the second biggest search engine behind Google). Be specific with your YouTube search criteria. For example, search on "Red Driving School" or "the AA driving instructor training". Have a good look around. Videos posted on YouTube by individuals are usually the most informative, as opposed to the corporate ones.

Next, do your market research for driving lessons in your own area. For example, if you live in Telford, do a Google search "driving lessons Telford".

Look for the local businesses, which appear on the first page of your search and click on the links to their websites.

For example, as I write this book, the website www.drivinglessons-telford.co.uk comes up in third place on the search results (below two paid-for ads). Click on this website and hit the tab 'prices'. Ignore any introductory offers and look for the price of a block-booking of 10 hours. In this case a block booking of 10 hours is £250, 20 hours is £500, 30 hours is £750 and 40 hours is £1,000. So they're not discounting at all, which indicates that demand for driving lessons in the Telford area (on first impression at least) is good.

Block-booking of 10 hours from other driving schools in the Telford area cost between £240 and £270, so £25 per hour is a sensible base figure to use for planning purposes.

3. Make Plans

You will need to write three separate plans:

a) Your training plan

b) Your profit and loss forecast

c) Your marketing plan

Most aspiring ADI's don't do this, which amazes me. I get ALL my instructor pupils to write their own training, financial and marketing plans, which I review with them in detail throughout their training period.

Let's have a look at what goes into each plan...

a) Training Plan

- Who will be your training organisation?
- Who will be your trainer for part 2 and part 3? (if your training organisation can't give you that information, the most likely reason is that they don't have anyone based in your area, so they will need to find someone and it's pot luck who you get)
- What is the availability of your trainer? (some trainers can be booked up for months ahead)
- Where will the training will take place? Is it close to you or will you have to drive for miles?
- Which test centre you will use for your part 2 and part 3 tests?
- What is the current waiting time for part 2 and part 3 tests?

The lead times for ADI part 2 and part 3 tests can be up to 15 weeks in some parts of the country, due to the availability of suitably qualified examiners, so it may be worthwhile looking at alternative test centres if you need to book an earlier test date.

Book your test dates at the earliest opportunity. Don't wait until you've finished your training. You're allowed to change your test centre and test date up to 6 times free of charge if you feel you're not ready.

Your training plan may be structured like this (enter your own expected completion dates)…

Activity	Expected Completion Date	Cost
Apply for DBS check	week 1	
Receive CRB reference number	week 3	
Get PRN number	week 5	
Book ADI Part 1 test (theory)	week 5	
Study for part 1	week 6- 7	
Take (and pass) Part 1 test	week 8	
Book ADI part 2 test	week 8	
Train for part 2	weeks 9-11	
Take (and pass) part 2 test	week 12	
Book ADI part 3 test	week 12	
Train for part 2	weeks 12-20	
Take (and pass) part 3 test	week 20	

b) Profit and Loss Forecast

Income

a) How many weeks per year will you work? (eg 44)

b) how many hours in a typical week (mornings, afternoons and evenings) will you allocate to driving instruction (eg Saturday 4, Sunday 0, Monday to Friday 6 = 34)

NB : The number of hours you may want to work per week may be different in summer and winter. If so, write an Excel spreadsheet by month

c) What is your hourly rate? (say £25)

Gross income = a) × b) × c)   so 44 × 34 × 25   **= £37,400**

Costs

a) Franchise fee (if applicable) per week (say £235) = £12,220 per annum

b) Car cost (if applicable, incl lease costs, insurance, MOT and servicing)

c) Fuel costs @ £2 per hour of driving (so 44 weeks × 5 days per week × 9 hours per day, including time in between lessons) = £3,960

d) Marketing costs including website, social media, local adverts, business cards, polo shirts, leaflets. Say £1,000

e) Budget on 5% of my lessons being cancelled due to bad weather, illness or car servicing (so total lost income at 5% of £37,400 is £1,870

f) Mobile phone @ £50 per month = £600 per annum

g) Accounting fees and tax returns £500 per annum

h) Other expenses (courses, ADI meetings, manuals, travel expenses etc) £500

Total costs (in this example)...

Franchise fees		£12,220
Fuel		£3,960
Marketing		£1,000
Lessons cancelled by me		£1,870
Mobile phone		£600
Accounting		£500
Others		£500
Total Costs	=	£20,650
So, net income before tax	=	**£37,400 - £20,650**
	=	**£16,750**

These figures (although made up) are very realistic of the true net income you can expect before tax. They are a long way short of the '£30,000 to £50,000' income figures you'll read about in the glossy brochures. These are your gross incomes, before deducting your costs.

So how do you increase your net income?

Either work longer hours and increase your income, or reduce your costs.

What is your main cost? The franchise fee.

This is the main reason why the majority of driving instructors who have been in the business for more that 3 years run their own independent driving school.

Marketing Plan

Where will you get new leads from?

Select three or four of the marketing ideas I've listed in the Marketing chapter of this book and develop them further. For example, if you're planning on advertising in local shops and post offices, decide which shops exactly and find out how much they charge.

4. Take the First Step

Once you've made a commitment to at least find out more, think positively in terms of 'how I'm going to make this happen'.

The first thing to do is to check that you're eligible…

Go to www.gov.uk/become-a-driving-instructor to check if you can become an approved driving instructor (ADI) and what you need to do to qualify. The .gov.uk website is really helpful and is easy to follow.

Assuming you're eligible to apply then the next step the .gov.uk website takes you through is a Discosure and Barring Service (DBS) check. This costs £6 and I would suggest you get on with it and do it straight away. It's only £6 and it's always useful to have. Just click on the link on the .gov.uk website and follow the instructions online.

The organisation PIN you will need to enter is '105205'

The secret word is 'axis'

The organisation name is 'DVSA – PDI (PO)'

Alternatively you can ring GBOnline on 0845 251 5000.

The DBS check usually takes around 2 weeks, so it's worth cracking on with.

Once you've received your Criminal Records Bureau (CRB) disclosure number you can then register as a PDI (potential driving instructor) and receive your Personal Reference Number, which you will need to book your ADI tests.

More importantly, it means you've taken your first steps and you're on your way. It will definitely save you a lot of time by doing this step as soon as possible.

It's no longer just wishful thinking: you're taking action.

5. Take Responsibility for your Training

This is a life changing opportunity and you really need to grasp hold of it and take responsibility for your training.

By this I mean if you're not happy about something then take positive action yourself to get it put right. I changed my training provider twice during my training period, because I was not at all satisfied with the quality of training and support I was being given. What I discovered was that the post-sales support my original training company actually provided during my training was woeful, compared to the bells and whistles approach given to me by their sales team. I was suckered into joining them and I fell for it. Lesson learned. Eventually I abandoned the training that I had paid for up-front and instead I had private lessons from the former head of examiner recruitment and training at the DVSA. The guy who had previously run the training courses to teach the part 2 and part 3 examiners!

He explained in detail exactly what the examiners were looking for. As a result of his training, I passed my third and final attempt at my ADI Part 3 test just one mark short of a perfect score. The instructor training I offer in my own courses is based on exactly the same material as I received from him.

Badger your training company and become a source of irritation if necessary. Buy additional training material from other providers, over and above what your own training organisation gives you.

I spent hours and hours watching training videos from www.driving-instructor.tv

There is a lot of repetition on the videos, but to me it didn't matter. It was great material, which I highly recommend to all my pupils preparing for their ADI part 3 test. But I only discovered this little gold mine of information by taking action and asking other instructors what they did and what really helped them in their training.

In other words, I took responsibility for passing my tests and I was not going to blame anyone else if I didn't pass on my third and final attempt. It was totally my responsibility to pass my part 3, and if I didn't then I only had myself to blame.

That's the approach I took and it worked for me. I urge you to do likewise.

6. Be Efficient - Keep your wasted time to an absolute minimum and be prepared to say 'No'

Once you're at the stage of giving lessons, I urge you to do what my partner made me do, which was to analyse how I was actually spending my time.

When I first started giving lessons, a typical day in my diary looked like this...

08:30 – 10:00	Jodie
10:30 – 12:00	Ethan
13:00 – 14:00	Donna. Lesson cancelled
14:30 – 16:00	Ben
16:30 – 18:00	Karen

So, that's 9.5 hours from collecting the first pupil to dropping off the last one.

In that time I gave a total of 6 hours of lessons. That's 63% efficiency. As all my lessons were pay-as-you-go, I didn't get paid for any lessons that were cancelled by my pupils.

Today my day typically looks like this…

08:30 – 12:30	Sophie
13:00 – 17:00	Tom

8.5 hours from start to finish. 8 hours of paid lessons. 94% efficiency. I finish an hour earlier and get paid for 8 hours as opposed to 6 (an increase of 33%). All my lessons are pre-paid, so if a pupil cancels at short notice then I try and fill the slot. If I can't fill it, then the cost of the lesson is deducted from the lump-sum the pupil has already paid.

Saying no occasionally is vital. For example, this actually happened to me recently:

A pupil of mine who is test-ready had his test cancelled by the DVSA and rescheduled for 5 weeks time. He asked me for an hour's lesson a week for the next 5 weeks from 5:30pm til 6:30pm after he finished work. To get to him at that time of day meant fighting my way across town during rush hour. Probably 30 minutes to get to him and 30 minutes to get back. My answer was no, but I didn't say that directly to him. Instead, I said that it would be best for him to get the maximum value from his lessons by driving in the same areas where he was most likely to go on his test. To do this, it would be better to do one 2 hour lesson rather than two 1 hour lessons, which he was happy with. I also suggested that we started later after the rush hour, which he thought was a good idea. So I got what I wanted (i.e. better efficiency) and at the same time kept him happy, as I had worded it in such a way that it was in his best interests to do it my way.

Be prepared to say no for other reasons as well:

I don't work on Sundays. Full stop.

I work a maximum of two evenings per week in summer and none in winter. Full stop.

I occasionally say no to some pupils after our initial lessons, if I really don't think it's going to work.

Ring fence key dates in your diary. If you want to play golf on a Friday afternoon then cross Friday afternoons out. Work around it. Likewise birthdays, school sports day, holidays, special anniversaries and so on.

When my diary is full, it's full. I have pupils on a waiting list. I don't try and fit them in on top of a full diary. You'll quickly become run-ragged if you say yes all the time.

I also insist on my pupils pre-paying for their lessons. The way I sell it to them is by saying that "I charge £30 per hour for lessons, but most pupils buy a block of 10 hours of lessons for £250, because it's a lot cheaper that way".

Nothing goes in my diary until it's paid for. That way I get very few last minute cancellations, and if a pupil does cancel at short notice then it's paid for it anyway.

Most of the driving tuition I do now is via intensive courses. Typically between 20 and 40 hours, all pre-paid. Four hours per day per pupil works best for me, as I can fit two pupils in per day and it's not too mentally demanding for the pupil.

I have tried 6 and 7 hour long lessons, but I've found that these are too long for both myself and for the pupil.

For me, the ideal working day is 2 x 4 hour lessons. That gives me a GROSS income of £200 per day (before costs) and it means that I'm not charging about all over the place. Happy days.

7. Sell driving courses, not driving lessons

Selling driving courses, as opposed to individual driving lessons, has worked really well since I started promoting them. It's much easier to get pupils to commit to buying blocks of lessons from me (which is what you want them to do) when their mindset is fixed on buying a driving course, as opposed to individual lessons. We cover exactly the same stuff, but it's just packaged differently and it works!

I offer three courses, each one merging into the next. Pupils can start at any level, as over half of the pupils who come to me have had some previous driving experience.

Here's what I include in each course:

Course 1 (New Driver)

Includes...

- Controls
- Moving off from the side of the road
- Stopping on the left
- MSM routine (mirror, signal, manoeuvre)
- Turning left at a junction
- Turning right at a junction
- Emerging left at a T junction
- Emerging right at a T junction
- Hill start

To complete the course, pupils have to drive around a route that I set, which incorporates around 20 junctions. I allow them no more than 3 faults (I'm pretty lenient with what I consider to be a fault). If they're successful then we move on to the next course. If not, then we practice whatever they're weak at and try again.

Pupils usually take between 5 and 10 hours of driving before they move on to the second course. They love it when you tell them that they've successfully completed the New Driver course and that they're now an Intermediate Driver, sometimes after as little as 6 hours of driving instruction. Their whole posture changes and they smile as you give them a "high-5".

Course 2 (Intermediate Driver)

Includes…

- Town centre driving
- Roundabouts (and mini-roundabouts)
- Meeting oncoming traffic
- Forward Planning
- Anticipation
- Road positioning
- Adequate clearance
- Pedestrian crossings
- Maintaining progress
- Straight line reversing

Pupils usually take between 8 and 10 hours to complete this course. Then they're ready to start preparing for their test.

Course 3 (Test Preparation)

Includes…

- Emergency stop
- Parallel parking
- Forward bay parking
- Reverse bay parking
- Stop on the right, reverse 2 car lengths and re-join the road
- Use of mirrors (re-cap)
- Commentary driving (this really works!)
- Country lane driving
- Using Satnav
- Motorway driving
- Test routes 1 to 10

I have made up 10 simulated test routes at each of the test centres I teach at (Eastbourne, Hastings and Folkestone). I take my pupils around these routes under test conditions, so that they have a good idea what to expect when they take their test for real. The more hours the pupils do on this section, the greater their chances of passing their driving test. Pupils usually spend between 10 and 15 hours completing this course.

The quickest any of my pupils has completed an intensive course with me from absolute scratch to passing his test is 13 hours of instruction over a 5 day period. He started with me on the Saturday and passed his test on the following Thursday. This is unusually quick though. Most pupils need between 30 and 40 hours in total. Less if they've ridden a motorbike before, as a lot of the skills (such as positioning at junctions) are transferable, and therefore they usually learn a lot quicker.

Think of it like this...

- A complete beginner pupil is worth (on average) £1,000 to your business over the course of all the time that they will spend with you
- A partly trained pupil is worth, on average, around £500 (depending on how many previous lessons they've had).

So look after them well and communicate with them frequently.

And as a bonus, tip 8 is:

8. Promote your own business, not someone else's

I see it every day. Cars painted from top to bottom with "AA", "RED", Bill Plant etc.

0800 contact telephone numbers, going through to a Head Office call centre.

I've got nothing against these organisations, by why oh why promote someone else's business when you could and should be promoting your own?

You want phone calls to come to YOU directly and not to a call centre, where anyone could benefit from any resulting new business. This is essential if you want to go down the route of taking on other drivers and building your own driving business.

You want YOUR business to become YOUR asset, which you can sell at some point if you choose to do so. That asset is only ever going to have a commercial value to you as the business owner if you create and build your own identity and attract new customers directly to you.

The alternative is that you promote and build someone else's business and all you end up with is a self-employed 'job' with no real value. Maybe that's all you're looking for and if so then that's fine. But I prefer to teach my instructor pupils to think with a business hat on and look at ways of not only generating a good income but also of building their own business, which will have a value.

Then I have to mention the complete un-readability of some roof boxes. For example:

Rob Brown 07856384284. Who on earth can remember a number like that in the 5 seconds or so that it takes to drive past you?

Much better to display something such as a website name TestSuccess.co.uk or

DriveHull 0800 100 9595. They're both easy to remember in the 5 seconds it takes to pass by you in the road. The quick test is to look at your wording for 5 seconds and then see if you can recall it. If you can remember what's written on it exactly, then that's a good sign. It's passed the 5 second test.

The best logo I've seen on an instructor's car is 'LeaP' with a red letter L and green letter P. Their slogan is "Jump from L to P. Call 0800 33 44 55". Great catchy logo that's easy to remember and a really strong call to action.

If you decide to take out a franchise with one of the large national companies, then be careful how you promote yourself. Read your contract well and if necessary obtain their permission first before having your own business cards and leaflets printed. Find out what you can and can't do regarding your own marketing literature.

I would be very wary of any organisation that doesn't allow you to promote your franchise in whatever way you want to. The question to ask yourself is whose interests do they care about most.... yours or theirs?

CHAPTER 10 – How Much Will I Earn?... and 30 Other Great Questions Answered

In this section, I've answered the 30 most common questions I've been asked by aspiring driving instructors.

1. How much will I earn a year?

This all depends on how many hours you work. Don't believe any of the headline figures you may see of £30,000 - £50,000 per annum. These are completely misleading. As a guide, if you give lessons for 6 hours a day, 210 days per annum at £25 per hour then your gross annual income will be around £31,500. But remember that you then have to deduct your franchise fees, fuel costs and tax. This brings the net annual income down from £31,500 to around £17,000 before tax, based on the same figures as above.

2. How much will my training cost?

Typically around £2,000 for the training itself. Plus there are test fees, licence fees and manuals. Budget on a total training cost of between £2,500 and £3,000. Beware of anyone offering cheaper training. It's usually linked to a 12-month or 24-month contract.

3. Can anyone become a driving instructor?

The rules state that you have to be aged 21 or over and have held a full car licence for at least 3 years. You have to pass a CRB check to check for any criminal convictions. Your application to become a driving instructor may be refused by the DVSA if you have more than 5 points on your licence, been banned from driving in the past or have been convicted of any non-motoring offences. There's a comprehensive list of what is required to be considered a "fit and proper person" on the .gov.uk website.

4. How long does it take to qualify?

You have to complete the process within 2 years from the date you pass your ADI part 1 theory test. The quickest any of my own pupils has qualified as an ADI is 5 months. He was fortunate to grab a last-minute test cancellation. The limiting factor is the availability of dates for the part 2 and part 3 exams. With some organisations it takes 12 to 18 months to qualify. People who train with me usually take between 6 and 9 months. If you can't afford to wait this long then contact me about my fast-track programme.

5. Will I be employed or self-employed once I complete my training?

Self-employed. It's your own business.

6. Am I guaranteed a job at the end of the training?

No. There's no job at the end of the training. People can be guaranteed a franchise, but that's not a job. You'll be self-employed.

7. Does everyone qualify?

No. The pass rate for the ADI part 3 exam in 2017 was a lowly 27%, and you only get 3 attempts. More than half of all pupils nationally who start their training either drop out or fail their part 3 test.

8. Is it mainly evening and weekend work?

No. You work the hours that suit you.

I try and make the most of the summer months and I do evening lessons 2 or 3 times a week months when it's still light. In contrast, in the dark winter months from November to January I usually finish around 4pm. I also work occasionally on Saturday mornings, but I do so because I choose not and not because I have to.

If a pupil works 9-5 Monday to Friday and only wants lessons in the evenings or at weekends then I rarely take them on. But that's my choice. You can work whatever hours suit you. You could even so the exact opposite to me and only work evenings and weekends. It's your choice.

9. What's a typical day for you?

In summer I typically collect my first pupil at 08:30am and drop off my last pupil at 18:30pm. So 10 hours elapsed from first pick-up to last drop-off. Sometimes I have a break from 3pm til 6pm and then go out again from 6pm til 9pm when it's still light. I really enjoy working in the evenings in summer, as the roads are so much quieter.

In winter, I usually collect my first pupil at 08:00 and finish at 16:00. I also do more Saturday morning work in the winter to make the most of the daylight. I don't enjoy teaching in the dark, but many other instructors don't mind at all. But that's their choice.

10. Is there sufficient demand to keep you busy all year round?

Yes. Certainly in my area along the south coast from Eastbourne to Hastings, there's plenty of work all year round. The busiest time of year is from early July to mid September. A-level pupils have finished their exams and teenagers are returning home from University for their summer break. They all want to pass their tests by mid Sept, so I'm usually working long hours to keep up with demand. I don't mind this at all. Make hay while the sun is shining!

11. Are examiners only allowed to pass a certain number of new instructors each year?

No. Definitely not.

12. Is it best for me to manage my own diary or have someone else manage it for me?

That's very much down to personal preference. The majority of driving instructors (myself included) manage our own diaries. I want complete control of my time.

13. Do pupils crash their instructor's car?

Occasionally yes, but it's rare. I've had one bump in the last 4 years when someone went into the back of me at around 5 mph. The most common damage is caused not by pupils' driving, but by pot holes in the road. I'm on my third set of shock absorbers in the last 2 years.

14. Is car insurance expensive?

It can be, but like all car insurance it depends on your age, no claims discount, location etc. As a guide, my car insurance costs me £405 per annum. The car insurance for a 24 year-old instructor I know costs her over £2,000 per annum. I suggest you get a couple of quotes as part of your own due diligence.

15. What's the best car to teach in?

One fitted with dual controls!

The big decision is whether to teach in a manual car or in an automatic. You can usually charge slightly more for driving lessons in an automatic car, but there more pupils looking to pass their test in a manual car, as this covers them for driving both manual and automatic cars.

Consider hybrid cars such as the Toyota Yaris. Hybrid cars can be a lot more economical to run around town. They run on both electricity and petrol.

Popular cars for instructors are Ford Fiesta, Vauxhall Corsa, Peugeot 108 and 208, Toyota Yaris, Citroen C1, Mini Cooper and the Renault Clio. Petrol or diesel. Usually 1.0 to 1.6 litre engine. The choice is yours. If you're joining a national franchise then your choice of car may be limited.

16. Should I use my own car to teach in or should I buy or lease a new one?

You can do either.

You can use your own car if it's suitable. I lease my driving instruction car and I also own an estate car, as I have two wonderful black labs.

You can lease cars from many different places. Main dealers for Ford, Peugeot etc often have really good deals.

Two companies that specialise in lease cars for driving instructors are

Hitachi	https://www.hitachicapitalvehiclesolutions.co.uk/driving-instructor-leasing
CA Cars	www.cacars.co.uk/

17. Do I need to set up a limited company?

No, there is no need to. Most driving instructors are self-employed. I suggest that

you don't spend too much time initially on matters like this – they can wait. Concentrate on your training first. Then take professional advice.

18. How many miles a year will I drive?

As a guide, between 20,000 and 30,000 miles. I'm full time and I did 27,000 miles last year.

19. If I'm charging pupils for lessons on a PDI licence before I've finished my exams, can I still charge the full amount?

Yes.

20. Do you help pupils pass their theory test?

No. There's no reason why you can't, but I don't. I leave them to do it themselves.

21. Why is the pass rate for the Part 3 exam so low?

Good question. In my opinion there are three main reasons....

a) The quality of the training given

b) The intensity at which the training is delivered (40 hours in one week is ridiculous)

c) Lack of practice in the real world using actual pupils.

22. Are there any areas of the UK saturated with driving instructors?

Not to my knowledge. More existing driving instructors are retiring or leaving the industry than joining it.

23. What does the long-term future hold?

The biggest threat to the industry in the long term is driverless cars. Big changes will inevitably take place as new driverless technology is introduced and the industry will need to adapt to this change. More and more cars will also become electric over the next 20 years.

I don't see much changing in the short and medium term. There is a constant stream of young people reaching their 17th birthday and wanting to learn to drive.

24. Can you have music on in the car?

Yes, of course. I have the radio on a lot. Especially Test Match Special in the summer.

25. If the car is caught speeding while a pupil is driving, who gets the ticket?

Whoever is driving at the time.

26. If I really don't get on with a pupil, or they don't get on with me, do I have to keep them on?

No. If it's just not working out, then you're under no obligation to keep teaching them and I think it's best to wish them well with their driving, but with another instructor. I say no to about 1 pupil in every 20. The most common reasons for me not continuing with a pupil is their unreliability or if they're only available when I don't want to work, such as on Sundays.

27. What happens if it's icy or snowy?

I adapt lessons accordingly and stick to roads I know have been gritted. If road conditions are particularly bad then I'll reschedule the lesson. That's why I like to work long hours in the summer months, so that I can allow for lesson cancellations in the depths of winter.

28. What happens if a pupil cancels their lesson at short notice?

You have to be firm but fair. I make my cancellation policy very clear from the outset and I state that I require at least 24 hours notice of cancellation. Virtually all of my lessons are pre-paid, so the pupil would lose their lesson if they cancelled at short notice. You really can't afford to be messed around by pupils, as you have a business to run.

Some of the reasons I've heard include:

- I woke up this morning not feeling well
- My child was sick in the night, so I need to stay at home to look after him/her

- My dog was sick in the night and I need to take it to the vets
- I've been called into work
- I must have eaten something dodgy last night
- I don't want to drive when it's raining
- It's a bit foggy outside
- I've had to get an emergency doctor's appointment and the only time they could fit me in is 10:30am
- I've got a washing machine being delivered today, so I need to be here when it arrives.

It's amazing how many pupils make incredibly quick recoveries or make alternative arrangements when they realise that they will lose their money if they cancel.

It doesn't happen that often, but when it does you need to be firm but fair.

29. Which area should I work?

You want to work in as tight an area as possible and as close to where you live as possible. Ideally in an area with a population of over 50,000 and with its own test centre. Ideally no more than a 15-minute drive between pupils.

My own area covers Eastbourne, Bexhill, St Leonards, Battle and Hastings. Total population is over 100,000 and there are test centres at Eastbourne and at Hastings. This works well for me. There's plenty of demand within a short radius, loads of variety and it's a pretty good area to work in.

If I get an enquiry from a pupil outside my area then I either say no, or I quote them a higher hourly rate to compensate for me having to drive further to collect them. I explain the reasons why I'm charging more. Some say OK and accept it. Others you don't hear from again, but that's fine with me. Don't chase after every enquiry. Be selective and take on whatever works for you.

30. Do I ever regret becoming a driving instructor?

No, it's been brilliant for me. The beauty of it is that it gives to total control of your life and the freedom to work as much or as little as you choose. You also get to meet new people all the time.

31. Would you have done things differently if you were starting out again?

Yes, most definitely!

The real issue most aspiring driving instructors have (myself included when I first started out) is that you're going into this industry fairly blind. You don't know what you don't know. It's like travelling to a remote island...you don't know what to expect until you actually get there. Hindsight is a wonderful thing. I would have done things very differently if I knew then what I know now. And that's the main reason for writing this book.... to help you guys.

Hope this helps. If you have any other questions, then please feel free to email me at

bc@briancroucher.co.uk. I'll do my best to answer them.

CHAPTER 11 – What's Next For You?

Right, so now you've read the book, what's next for you?

Whether you're looking at becoming a driving instructor on a full-time or on a part-time basis, there are several things you can get started on right now.

Please don't just put this book back on the shelf and think "I'll do it one day". Instead, keep the momentum going and take the next steps to find out more.

Here's what I suggest you do:

1. DBS Check

Go ahead and get your DBS check carried out (as explained in the 7 Keys to Success). It only costs £6, so just get it done and out of the way. It'll certainly save you time in the long run.

2. Get Information Packs and find out the availability of training

Start collecting the brochures from the main players of the industry and have a read through. Start comparing one against another (The AA, Red Driving, Bill Plant etc). Phone them up. Have a chat with a representative from their sales team. Find out the cost of their training, how long it normally takes to complete their training program, when their next course starts and whether or not you're tied in to a contract at the end of the training. If so, for how long?

3. Come on a 'Test Drive Day' and experience what it's like

Would you buy a new car without going on a test drive first? So how about coming on one of my 'test drive days'? You get to come down to the south coast and join me and one (or sometimes two) of my pupils for half a day and experience what it's really like to be a driving instructor. You'll sit in the back of my car on a driving lesson (or lessons) for half a day. These half days can be either in the morning or in the afternoon to suit you and it usually includes lunch either before or after. Importantly, you get a chance to have all your questions answered. There's a nominal charge of £40 for this, but I deduct this from the cost of your training should you decide to go ahead. Many people who have come on such test-drive days in the past have found them to be a real eye-opener.

As I'm sure you've gathered, I run my own driving instructor training business. It's a relatively small operation, but pupils find that being small has big advantages, which work really well.

To check availability of spaces on my up-coming ADI training courses, email me at bc@briancroucher.co.uk

The courses are particularly popular with my pupils for the following reasons:

a. Flexibility

Start when it suits you. I don't have any rigid course dates, as I only usually take on two new pupils each month. The training can be an intensive course, or it can be spread out over a longer period of time. Whatever works for you. We can do some of the training in the evenings and at weekends if necessary.

b. Speed

Many of my pupils want to start earning money asap, so we create our own fast-track path, tailored to the needs of each pupil. This enables them to start earning in less than half the normal time it takes with other training companies. The biggest time consumer is not the training, but the availability of test dates.

c. Freedom

There's no contractual tie-in after you qualify. That is so important. You're free to go down whichever route you choose. I go through all the options with you and help steer you in the direction that best meets your own particular needs.

d. Cost-effectiveness

You only pay 25% of the course fees to reserve you place. This covers your training for both your ADI part 1 test and your ADI part 2 test. You then decide whether you want to continue with me and train for your ADI part 3 test. Most pupils do, but not everyone. And that's fine. Much better to find out at this stage before you have to pay large amounts of training fees.

Also, as a thank you for buying this book, there's a £100 training voucher included. Just quote '**VOUCHER100**' when booking your training.

e. Success rate

As I normally only take on two new pupils per month, the training is very much geared to your own specific needs, as opposed to having a 'one size fits all' approach. Unsurprisingly, our pass rates are well above average.

And last, but certainly not least....

f. Become a Regional Instructor Trainer

Complete your ADI training with us, qualify as an ADI and gain a minimum of 12 month's practical experience as a qualified ADI instructor. Then consider boosting your income by becoming one of our Regional ADI Instructor Trainers. A great additional arm to your driving business. The hourly rate is normally 20% higher than for normal driving lessons and we supply you with pupils in your area.

We only take on pupils as Regional Instructors Trainers who have completed their ADI training with us, as we want you to be able to share exactly the same experience. We're looking to recruit up to four Regional Trainers in each of the following regions...

- Scotland
- Wales
- Northern Ireland
- North-west England
- North-east England
- North Yorkshire & Humberside
- East Midlands
- West Midlands
- Eastern England

- South-west England
- Greater London
- South-east England.

If this is of interest then I would like to hear from you. Please email me at bc@briancroucher.co.uk for further details.

Whichever route you decide to go down with your new career, I wish you every success on your own journey.

And as a final thank you for buying this book, here's your £100 money-off voucher:

£100 TRAINING DISCOUNT VOUCHER

QUOTE 'VOUCHER100'

TO RECEIVE £100 OFF THE COST OF YOUR ADI PART 3 TRAINING

And so, on to the next 100 test passes.

Congratulations Liam on being number 101!

Printed in Great Britain
by Amazon

65555633R00077

FIELD NATURALIST'S LIBRARY

Fungi

OF BRITAIN AND NORTHERN EUROPE

FIELD NATURALIST'S LIBRARY

Fungi

OF BRITAIN AND NORTHERN EUROPE

PAUL STERRY
INTRODUCED BY TONY SOPER

Bounty Books

First published in 1991 by Hamlyn, a division of Octopus Publishing Group

This edition published in 2011 by Bounty Books,
a division of Octopus Publishing Group,
Endeavour House,
189 Shaftesbury Avenue
London WC2H 8JY
www.octopusbooks.co.uk

An Hachette UK Company
www.hachette.co.uk

ISBN: 978-0-753721-51-3

Artwork by Sean Milne ©Equinox Limited 1979
First published in *Mushrooms and Toadstools, a field guide* Oxford University Press 1979

Printed and bound in China

Contents

Introduction by Tony Soper

Fungi have never had the instant popularity of warm-blooded furry mammals with big round eyes, or captured our imagination as does the soaring flight and colourful display of birds. On the face of it they are anaemic organisms, lacking the chlorophyll which gives that healthy greenness to 'proper' plants. They include the various moulds which plague us in so many tiresome ways, singularly lacking a vital spark which brings endearment.

They are mostly seen in the light of problems. Not so long ago stored fruit, jam and meats were likely to sport a noisome slime which reduced their eye-appeal only too effectively. The long fight against moulds encouraged us to develop storage methods which began with brine solutions and followed through to present-day techniques of sterilisation and freeze-drying. Damp wallpaper peels and turns black with *Cladosporium*, and the less said about dry rot the better. The devastating power of fungal mould is only too familiar to all of us. Yet one of our great blessings, penicillin, is derived from a common green mould, and doubtless many others are yet to be discovered.

Fungus is a doleful word, and only too many of our encounters with it are uncomfortably mysterious, if not painful. Tombstones rise from the warm damp grave, lifted by an irresistable and long-dormant force. Paving slabs suddenly rear up overnight to trip the unwary passer-by. Fungi do their dirty work in dark and secret places, reaching along the roots of a seemingly-healthy tree, only to strike it down while it seems at the height of its maturity. It seems that to the average Briton the only good fungus is the cultivated mushroom or the field mushroom fried with bacon and eggs.

There are very few common names for fungi, a sure sign of general ignorance and indifference. Most of us classify them as either mushrooms or toadstools, a recipe for muddle and confusion which this book seeks to resolve. The general belief that all 'toadstools' are poisonous and only 'mushrooms' are safe has led to a depressing number of violent stomach aches and an alarming number of deaths. The truth is that there is no easy way to tell when a 'mushroom' is safe to eat. All the rule-of-thumb methods and all of the old-wives-tales are best forgotten. There is no substitute for careful study of characteristics and habit.

Field mushrooms peel easily, and that is a character commonly regarded as a test of edibility. But the Death Cap, *Amanita phalloides*, peels easily, too, and repays those who eat it with a drawn-out and highly unpleasant demise. The old-wives tale was that provided a silver spoon didn't turn black in the cooking mushrooms, all was well. But *Amanita phalloides* evades that test as well. So be careful, but remember that the number of poisonous fungi is in fact very small. Study the book well, and stay on the straight and narrow path!

The whole history of fungology, or mycology as the authorities have it nowadays, is fascinating. From the days when it was believed that these organisms were born of an excrescence created by the roar of thunder to the slow realisation that they reproduce by spores and not by seeding, the struggle to introduce fungi to a more enlightened public has been an uphill one. Yet we would be in trouble without their invaluable help. As scavengers, they clean up rottng and dead organic material to recycle it into productive soil. Without their actions, we would all be buried under a monstrous mound of corpses.

But put aside the morbid thoughts which always spring to mind at the mention of fungus and mould, and give them honest consideration. They are not only beautiful, in many cases they are endlessly interesting.

INTRODUCTION

Fungi are an intruiging and fascinating group to study. They are extremely varied in their shape, structure, and colour and most have unusual methods of feeding. Indeed, the way in which fungi obtain their energy has earned them a place in a kingdom distinct and separate from both plants and animals.

Fungi can be found in almost every habitat. Of the many thousands of species recorded in our region, most are too small or insignificant to be of interest. However, this still leaves nearly a thousand species of larger fungi to look for. Of these, the 250 or so species covered in this book are commonly found and reference to allied species in the text allows unusual ones to be covered as well.

Most people's image of a fungus is that of a familiar mushroom or toadstool. In fact these structures, important though they are, are just the reproductive or fruit bodies and occur only once a year. Most of the persistent part of the fungus remains hidden from view, in the soil or among the feeding substrate. The 'body' of the fungus comprise vast numbers of thin, branching filaments known as hyphae. These permeate the growing medium, forming a matted web called the mycelium. Except where densely packed, the mycelium and hyphae are too small to be seen by the naked eye.

Many fungi bear a superficial resemblance to plants in both shape and structure. However, they lack the green pigment, chlorophyll, which enables plants to manufacture most of their own food from sunlight, carbon dioxide and water. Instead, fungi feed more like animals relying on the decaying, and in some species living, tissue of plants and animals as a source of nutrients. Saprophytic species – those which use dead organic matter – are important in the recyling of plant and animal matter into the food chain. Parasitic species – those which attack living plants and animals – are significant factors in regulating numbers of plants and animals and some cause large losses in Man's crops and livestock.

Although many species of fungi attack and damage trees and plants,

Fungal mycelium on rotting wood

Honey Fungus growing on old birch

there are some plant-fungus relationships which are more benign. In a few, the mycelium grows in close association with the root system of the host – a symbiotic partnership in which both parties benefit – each obtaining nutrients from the other. Host plants with a mycorrhizal association tend to fare better than those without. Many orchids go one stage further and cannot germinate or grow successfully without the correct fungal partner.

While growing and feeding, fungi are essentially sedentary organisms. In order to disperse and colonise new habitats and food sources, they produce microscopic spores. In typical mushrooms and toadstools, these are formed on gills or in tubes. The spores are generally between 0.001 and 0.002 millimetres in diameter. In many, but by no means all, species of larger fungi the spores are carried on the wind. Even though the vast majority fail to form new mycelial colonies, the production of countless millions of spores by each fruit body ensures that a few survive.

The terms 'mushroom' and 'toadstool' are widely used but may mean different things to different people. To some a 'mushroom' is simply an edible species of toadstool while to others it may comprise members of the genus *Agaricus* – the familiar mushrooms from which the cultivated form is derived. The term 'toadstool' is sometimes used to describe all larger fungi while it may also be applied to inedible or poisonous species.

To the amateur naturalist, fungi are an interesting source of study on many levels. For some, searching for and finding some of our more bizarre and colourful species is reward enough in itself. For those who enjoy a challenge, the identification of some members of the larger fungus families may require keen regard to details and undaunted persistance. Many species make interesting – and static – photographic subjects but as a final reward after a hard day's foray, a few species are delicious to eat. With the exception of some of our more threatened species, the picking and eating of the occasional specimen will do little harm to their populations.

THE STRUCTURE AND CLASSIFICATION OF LARGER FUNGI

The classification of fungi is complex and the kingdom is now divided into several subdivisions. Of these, only the Ascomycotina – the Ascomyctes – and the Basidiomycotina – the Basidiomycetes – include species which grow to sufficient size to be recognised as larger fungi and considered in this book. Although members of the two subdivisions are often radically different in appearance, their means of spore production is the character used to define and distinguish them.

Ascomycetes

Fungi which comprise this group are known as 'spore-shooters' because of the means by which the spores are liberated. Microscopic spores are produced within specialised, elongated cells known as asci, eight spores per ascus. As they mature, fluid pressure builds up and eventually ruptures the tips of the asci thus releasing the spores. These are ejected with some considerable force over a distance of several millimetres, a fact that can sometimes be witnessed if a mature and undisturbed Ascomycete fruit body is gently tapped. In some Ascomycetes, the pressure build-up causes a lid, called an operculum, to flip open allowing the spores to be violently ejected. The way in which the ascus itself is produced and the structure on which it is carried varies throughout the Ascomycetes and gives rise to further sub-groups, of which the Discomycetes and the Pyrenomycetes are considered in this book.

Discomycetes This comprises the Cup Fungi, Morels, Helvellas, Earthtongues and Truffles. The asci are found closely-packed and form a smooth layer called the hymenium which lines the inner surface of a cup-shaped disc. In Morels, this structure is highly modified and has become globular and convoluted, being carried on a stem. The asci line the cavities in the 'head' of the fruit body. In the Truffles, or Tuberales, the hymenium has become so convoluted that it is entirely enclosed within the underground body of the fungus. Spores are dispersed when the Truffle is eaten by an animal.

A Discomycete:
Bladder Elf-cup
(*Peziza vesiculosa*)

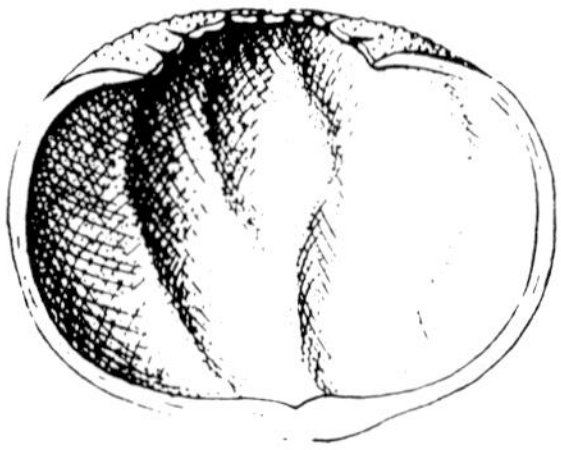

Longitudinal section of fruit-body

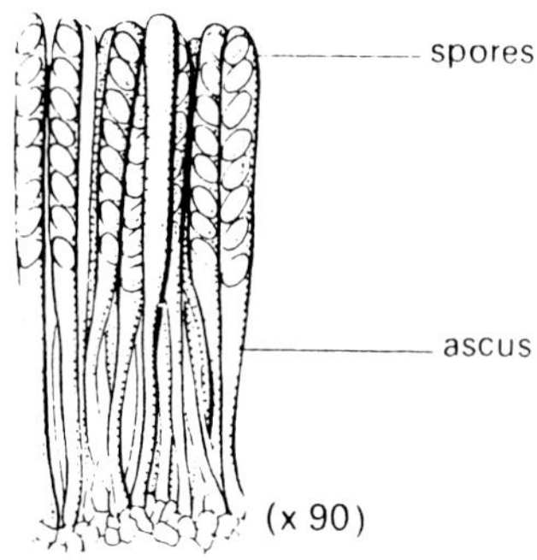

Part of inner surface

Pyrenomycetes This is a group known as the Flask Fungi because the asci are formed in flask-like structures. Of the species covered in this book some are small, colonial and pustular while others are hard and woody. Larger species in the region include the Candlesnuff Fungus, Dead Man's Fingers and King Alfred's Cakes.

Flask or pear-shaped structures, called perithecia, line the surface of the fruit body and contain the asci. The asci are orientated in such a way that they liberate their spores through the entrance pores to the perithecia.

A cross-section through a fruit body will reveal that the outer surface is lined with perithecia. In some species, the pores at the mouths of the perithecia can be seen on the outer surface of the fruit body with a hand lens.

For evidence that the Ascomycetes are 'spore-shooting' fungi, place a mature fruit body on a sheet of paper, cover it with a glass jar and leave it overnight. The result will be a halo of spores surrounding the fungus which have been expelled by force.

A Pyrenomycete: Dead Man's Fingers (*Xylaria polymorpha*)

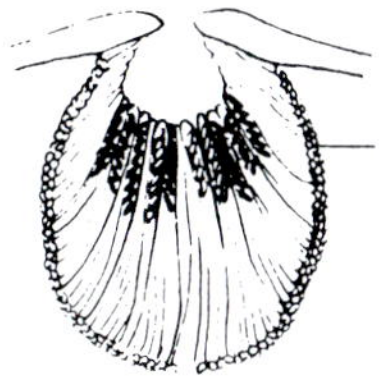

Longitudinal section of perithecium

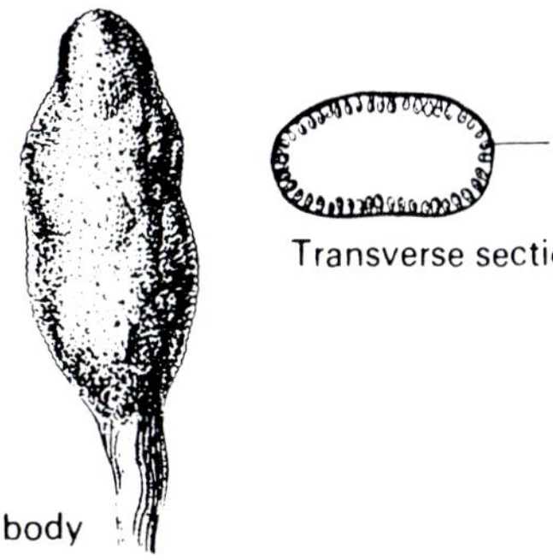

Orange Peel Fungus, Aleuria aurantia

Basidiomycetes

Members of this diverse group of fungi are extremely varied in shape and structure. However, even the most extreme forms have one thing in common, namely the way in which their spores are produced.

The Basidiomycetes are an important group of fungi. Most of the species of larger fungi found in the region belong to the Basidiomycetes and they comprise the majority of the species covered in this book. They include all the 'typical' gill mushrooms and toadstools, together with the Boletes, Bracket fungi, Puffballs, Earthstars and Stinkhorns.

In contrast to the other important group of larger fungi – the Ascomycetes – in the Basidiomycetes, the spores are produced by specialised, club-shaped cells called basidia which give the group its name. They are found on the outer surface of the fruit body but are often associated with highly specialised structures such as gill or pores.

When the fruit body matures, the spores, or basidiospores as they are more correctly termed, are ejected violently. However, because spore production may occur in localised sites on the fruit body, their dispersal further afield may require the assistance of factors such as wind and rain.

In the gill fungi, or Agarics, the surface area devoted to spore production is greatly increased by the presence of the closely-packed gill surfaces. These lie on the underside of the cap and are arranged radially from the apex of the stem. A section through the gills reveals that they are covered with a fertile layer of basidia cells called the hymenium. Spores are ejected from the basidia on the gill surface but they rely on the wind to disperse them further. By having the cap and gills raised off the ground by a stem, the spores stand a greater chance of catching a breeze. Not surprisingly, many of the spores fail to land on a suitable substrate and never grow to form a mycelial colony. However, vast numbers are produced in most species – it would not be unusual for in excess of 10,000 million to be produced by a single gill-fungus fruit body – and so it is almost certain that a few will survive.

In the Boletes, the gills are replaced

Fruit-body with vertical slice cut from cap margin

Gills cut across their length

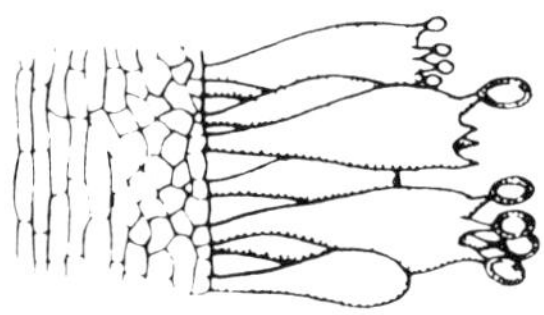

Basidia with spores

A Basidiomycete:
Field Mushroom
(*Agaricus campestris*)

Dryad's Saddle, Polyporus squamosus

by pores which serve the same function of increasing the surface area. If the cap of a Bolete or an Agaric is placed on a sheet of paper, covered and left for a few hours, a perfect impression of the gills can be seen. The spore colour is also easy to distinguish.

In both the Agarics and the Boletes, the colour and shape of the gills is an important feature used in identification. The spore colour is also useful for many common species and the spore shape, when viewed under a microscope, also helps experts distinguish closely related species.

In a few species of Basidiomycetes, the under surface of the cap is covered in spikes or pegs. These greatly increase the surface area for spore production.

In the Puffballs, which are members of the Gasteromycetales, the spores are produced in a large mass within the interior of the fruit body. Initially this mass is solid but with maturity, the spores become powdery. Before they can dipserse, however, the outer skin must first split to release them. In some species a neat pore forms while in others, the whole skin ruptures. The action of wind, heavy rain and being knocked helps liberate the spores which are then carried by the wind.

In the Stinkhorns – relatives of the Puffballs – the spore mass is carried on the head of the fruit body. A strong and unpleasant smell attracts insects such as flies which eat the spore mass and disperse it to new areas.

In the Agarcis and bracket fungi, the basidial cell bears two or four stalks, each of which eventually produces a spore. In another group of Basidiomycetes, the Jelly Fungi, the basidial cell is divided into four, and each section produces a spore-bearing stalk. The Jelly Fungi are an aptly-named group with soft, jelly-like tissues but are rather varied in appearance. Members of the group include the Jew's Ear, Yellow Brain Fungus and the Yellow Stagshorn Fungus.

IDENTIFICATION FEATURES

Fortunately for the amateur naturalist, several of the larger species of fungi found in Britain have either a unique appearance or a distinctive characters which make them easy to identify. However, many of the remaining species can be more troublesome. Some of the families appear superficially similar and the larger genera often comprise members that are closely related and look much alike. A beginner's first few fungus forays may be bewildering. However, paying close attention to certain key features as well as to the environment in which the specimen is found, can greatly improve the likelihood of successful identification.

Features to look for when identifying the larger Agarics include the shape and colour of the cap, gills and stems. However, the way in which the fruit body develops prior to maturity also adds some vital clues. A knowledge of the structure and the developmental stages can, therefore, be useful to give a background understanding.

The early stages of some Agaric fruit bodies are completely enclosed in a membrane called the universal veil. As the fungus expands, this is ruptured and the presence or absence of veil remains on the cap and a sac-like volva surrounding the stem base are important. As the cap expands it may leave a membranous ring attached to the stem in some genera or species. In those which bear a stem ring, this can be difficult to see or may be lost easily.

The caps of different Agarics and Boletes vary considerably and even individual specimens may change shape markedly as they grow. Mushrooms and toadstools generally have rather domed caps in their early stages. These may, however, expand and flatten with age. In some species the mature cap becomes inverted and funnel-shaped while others assume a flattened dome-shape. The centre of the cap may be raised in a few species; this area is known as the umbo.

Cap colour is also important. It may be a uniform colour, blotched or even zoned. Some species bruise a different colour and others change as the cap dries out.

The texture of the cap may also give

The fruit-bodies of agarics cut lengthwise to show their veils.

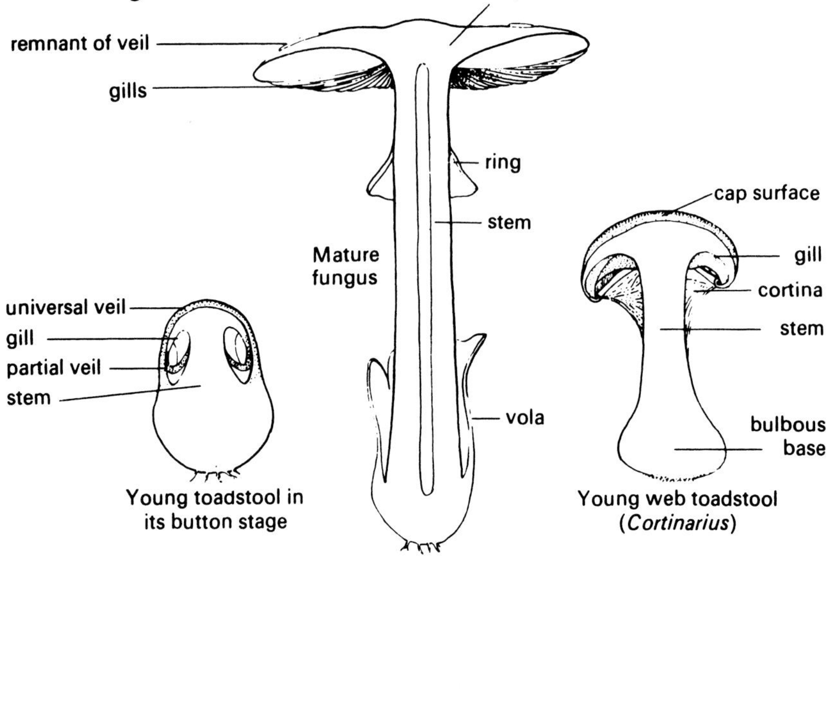

Fairies' Bonnets, Coprinus disseminatus

vital clues: some are smooth and dry while other species are sticky or covered with slime.

The gills on the underside of the cap should be studied closely. Note their colour, whether they bruise and their shape. Decurrent gills run down the stem while others are not joined to the stem. Some species have closely packed gills while in others they are widely separated. The spores may also have a distinctive colour. Sometimes this is visible on the gills but otherwise a spore print, performed at home, can be very useful.

Stems vary greatly from species to species. Some are straight while others are twisted and may have a bulbous base. Some are smooth while others are sticky or are covered in scales or hairs. The flesh within the stem may be hollow in some species and in others, several stems are fused together near the base. The flesh of the stem may be brittle and easily broken or tough and leathery.

The flesh of the fungus is also important. The colour and ability to bruise is often diagnostic and some fungi, most notably the in the genus *Lactarius*, produce a characteristic milk when damaged. The flesh may have a characteristic and diagnostic smell and taste. However, if there is any suspicion that the specimen may resemble a poisonous species, refrain from tasting it. Having noted all the important structural features of the fungus itself, the naturalist should also pay some attention to its envirnoment. The habitat is often quite specific. For example, some species will only grow in Beech woodlands on chalk soil while others only grow in association with certain species of trees. Some are always found on rotting wood while others prefer leaf litter. Noting this information provides further vital evidence to the specimen's identity. Lastly, consider the time of year. Most of the larger mushrooms and toadstools are extremely seasonal. Although only general guidelines, the periods of occurence indicated beside each species in this book, may help clinch an identification.

COLLECTING AND STUDYING FUNGI

Mushrooms and toadstools are a fascinating group to study. Some of our species are large and striking and easy to find. Others may require a persistant searching and more than a little luck. Careful preparation prior to a fungus foray combined with a little insight into the habitat requirements and peculiarities of some of our more unusual fungi, can greatly improve the chances of an increased species list for a day's outing.

Choosing a good destination for a fungus foray is an important decision. Although many of our fungi prefer open habitats and pastures, for sheer variety, woodlands offer the best opportunities for the more serious mycologist. Not surprisingly, the type of woodland has a great bearing on what will be found. Old, mature forest generally encourages a rich growth of mushrooms and toadstools, especially where fallen, dead wood and partly buried timber is left on the forest floor. Younger, more disturbed woods generally have less to offer, and monoculture plantations of conifers are usually extremely poor until they are at least ten years old. Although some of the larger species of fungi may be conspicuous, many match the colours of the woodland floor and may be partly covered in leaf litter. Try lying flat on the ground and scanning around you to improve the chances of finding the more delicate and cryptic species.

Mushrooms, toadstools and the other fruit bodies covered in this book are generally seasonal in appearance although some are more persistant than others. A few species may occur at any time of the year but most are found for only a few weeks or months, the autumn generally being the best season for variety. Because a high percentage of the fungus body is water, dry seasons are often less productive than wet ones. Try searching your local woodland a few days after the first heavy downpour of autumn, especially if the weather has been warm.

Fortunately for the beginner, many of the larger fungi are distinctive enough for identification to be made in the field. However, for a few speci-

The Sickener, Russula emetica

mens, it may be necessary to pick them and study them in more detail at home. When you pick a mushroom or toadstool, try to avoid crushing or damaging it too much and be sure to take the whole of the stem: for example, the presence or absence of a sac-like volva can be critical to the identification of some species. A flat, fruit basket is the best container to use for transporting fungi. Be sure not to squash specimens together and never store them in airtight vessels as many species will liquify rapidly as the moisture builds up.

Even if you decide to take a specimen home it is useful to make a few notes and observations *in situ*. Details such as the surrounding vegetation, nearby tree species, characteristic appearance (for example, whether it is sticky or dry), smell and taste may be vital in identification. These features may be lost after picking.

You may wish to make a spore print of the mushroom or toadstool when you get home either to assess spore colour or to make a permanant record. Cut the stem off neatly below the cap and place it, gills facing downwards, on a piece of paper. Leave it covered overnight and them remove the cap to reveal the print. To make the print permanant, cover the it with a sheet of sticky, transparent plastic.

Although picking mushrooms and toadstools generally does little to harm the mycelial colony of the fungus, it does leave fewer specimens for others to see and inevitably reduces, albeit slightly, the numbers of spores released into its growing habitat. Try to keep your collecting to the minimum required for identification and be realistic about the numbers you can cope with.

Another permanent and enjoyable way of recording your finds is to photograph them *in situ*. Most species look best when photographed at eye-level and so a tripod which can reach ground level may be useful. Larger species may need only a standard lens but for the smaller species, try extension rings or a specialised macro lens. Some people even fit extension rings to a wide-angle lens to give an impression of the mushroom's habitat. Some of the larger species of fungi have broad caps which cast a dense shadow beneath them. This can create problems for the photographer trying to capture the colour and shape of the gills. A useful method of overcoming this is to take a small, plastic mirror with you on every trip. By carefully positioning the reflective surface you can greatly reduce the effect of shadow and the contrast on the gills.

Be careful to keep good notes and records of what you photograph because identifying an unknown species from a picture is much harder than you might imagine. Make a note immediately after you have photographed each specimen.

Lastly, one of the most fulfilling aspects of going on fungus forays is the discovery of species new to the searcher. By revisiting a good site, year after year and indeed week after week in the autumn, not only are the best sites located but you can witness any seasonal or longer-term trends in fungus populations. Knowing an area well-enough to be able to predict a species' occurrence is extremely satisfying. It also heightens the delight upon discovering new and unexpected finds.

Magpie Cap, Coprinus picaceus

THE IMPORTANCE OF FUNGI

Their Significance to Man

Members of fungal world are varied, not only in their appearance and structure but also in their modes of life. While the vast majority of species may pass unnoticed and seemingly be of no significance to Man, some species are extremely important and a few have a profound effect on our everyday lives. Some fungi serve as food or influence our diet while others cause diseases to ourselves and to our crops and livestock. Far from just being fascinating curiosities, many fungi are of major economic importance.

The best known species of edible fungus in Britain and northern Europe is the cultivated mushroom, in fact a descendent of a species of *Agaricus*. Several of its wild relatives are enthusiastically harvested but regrettably their seasonal appearance is short. Continental Europe has broader tastes with regard to the species eaten. The Cep or Penny Bun, *Boletus edulis*, is a favourite as are the Chanterelle, *Cantharellus cibarius*, the Parasol Mushroom, *Macrolepiota procera*, and the Blewits, *Lepista nuda*. However, pride of place, both in terms of flavour and the price it commands, must go to the Truffle, *Tuber aestivum*, which in fungal terms is worth its weight in gold.

Although only distantly related to the larger mushrooms and toadstools featured in this book, the yeasts are nevertheless still fungi. Their role both in the diet and culture of European nations needs little comment. Their ability to produce carbon dioxide as a metabolic by-product in the presence of oxygen, and alcohol in its absence, has been harnessed by bakers and brewers for centuries. Yeasts are also a source of B-group vitamins, proteins and amino acids.

Fungi are also used to flavour cheeses and with many blue cheeses, one of several species of *Penicillium* may be involved producing unmistakable aromas and tastes. However, certain species of *Pencillium* have a different and more profound function as sources of Penicillin, an anitbiotic drug which has dramatically improved our ability to fight bacterial infections.

There are many species of fungi which definitely are not beneficial to

Burnt Polypore, Bjerkandera adusta, on an old oak stump

man. Cereal crops and fruit trees, for example, are subject to a wide range of fungal attacks from rusts and smuts which cause large-scale losses if not treated with fungicides.

The notorious Ergot fungus, *Claviceps purpurea,* which attacks cereal crops and contains alkaloids similar to LSD, not only reduces yield but is also extremely dangerous if eaten by humans or domesticated animals. A few species of larger fungi found in the region are also potentially dangerous. Most renowned of these is the Death Cap, *Amanita phalloides,* but several others such as the Destroying Angel, *Amanita virosa,* the Fly Agaric, *Amanita muscaria,* and the Red-staining Inocybe, *Inocybe patouillardii,* should also be avoided at all cost. Before considering eating any fungus you should make careful reference to the species notes in this and other books to be certain of its identity. If in any doubt about the identity or edibility of a specimen, always avoid it.

Their Role in Ecology

In addition to the important role that some species of fungi serve as sources of food and medicine for Man, the vast majority perform a far more fundamental and vital role in maintaining a healthy environment, and one which, by and large, goes unnoticed. As organisms that attack and break-down dead and dying organic matter, they fulfil a vital function in recyling nutrients back into food-chains. Were it not for the action of fungi, and, it must be said, bacteria as well, the energy and nutrients in biological systems would forever be locked out of the food chain.

To see the process of fungal break-down in action you need look no further than your kitchen. A mouldy loaf or pot of jam covered with a layer of *Pencillium* mould testifies to their rapid action. However, to witness the role of fungi in ecology in the more general sense, you should visit your nearest woodland. Rotting stumps and fallen branches are broken down over a period of a few years and the nutrients they contained returned to the soil to encourage new plant growth. Remember that mushrooms, toadstools and other fruit bodies are just temporary reproductive structures and that the main mycelial 'body' of the fungus is

Red Slug feeding on Amethyst Deceiver, Laccaria amethystea

at work, hidden from view in the soil and in the timber. Fallen leaves also become quickly incorporated into the leaf mould and again 'compost' the soil for future generations of plants and animals.

Fungi may seem so hardy as to warrant little in the way of attention from conservationists. Certainly, the pest species need little encouragement to survive. However, given the important recycling role that many species play in ecology, perhaps we should be concerned for the future of some of our mushrooms and toadstools. The biggest threat comes from habitat loss or, more insidiously, from habitat degradation. The over-enthusiastic removal of dead timber from woodland is one example of this and one which deprives many species of their prime habitat. With a knowledge of their vital role in decay, it is easy to see that the loss of fungal species is not only a loss in the visual sense but also a loss to the environment as a whole.

Amanita excelsia

GILL FUNGI

Many of the mushrooms and toadstools with which we are familiar have one feature in common – gills. These structures, on which the spores are produced, lie on the underside of a distinct cap. The gills greatly increase the surface area for spore production and their dispersal is assisted by being raised from the ground on a stem. Although almost all the species of gill fungi covered in this book have a clearly recognisable stem, cap and gills, there is still amazing variation in form, size and colour. Some, like the Parasol Mushroom, *Macrolepiota procera*, are tall and stout while others, like species of *Mycena*, are small and delicate. The gill fungi include some of the most delicious species of all but also a few of the most deadly.

FLY AGARIC
Amanita muscaria

Cap diameter: up to 15cm
Height: up to 18cm
Characteristics: One of our best known fungi. When fresh, the bright red or orange cap is dotted with white flecks. These are often washed off by rain. The stem, ring and gills are white and the bulbous base to the stem is ringed with scales. Spores white. Poisonous and hallucinogenic.
Range and habitat: Grows mainly in association with Birch and pine trees and is often found in mixed woodland and on heathland margins. Occurs throughout most of Britain and northern Europe where conditions are suitable. In good seasons, often forms extensive groups comprising fungi at different stages of development.
Similar species: Mature, undamaged specimens are unmistakable. Faded Fly Agarics may resemble *Amanita pantherina.* Rain-washed specimens which have lost the white flecks on the cap may resemble species of *Russula.* Check for the presence of the bulbous base and ring.

PANTHER CAP
Amanita pantherina
Cap diameter: up to 10cm
Height: up to 11cm
Characteristics: A robust species with a coffee brown or ochre coloured cap. When fresh, this is covered with white flecks or warts which are the remains of the veil. The gills are white and free. Spores white. The flesh is white and does not discolour when bruised or cut. The stem is stout and supports a ring. The stem thickens towards the base which is bulbous and is made more noticeable by the presence of pronounced rim. Above this, there are 2 or 3 membranous rings. Extremely poisonous.
Range and habitat: Widespread in northern Europe but rather local in Britain. The Panther Cap grows mainly under deciduous trees and is most frequently found under Beech.
Similar species: The Panther Cap may resemble *A. rubescens* whose flesh turns pink when damaged. Some specimens are similar to *A. excelsa* which lacks the Panther Cap's rim to the bulbous base.

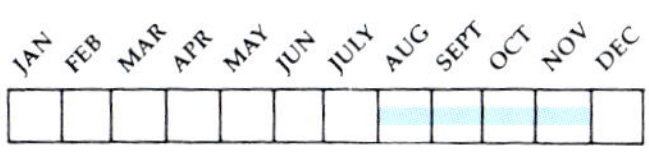

Fly Agaric, Amanita muscaria

DEATH CAP

Amanita phalloides

Cap diameter: up to 10cm
Height: up to 12cm
Characteristics: One of our most notorious fungi. The general colour of the cap and stem is yellowish-olive and lines radiate from the centre of the cap. Some specimens acquire a greenish tint giving them a sickly appearance. In damp weather, the cap is shiny. The gills are white. Spores white. The stem has a membranous ring and a bulbous base with a white, sac-like volva. The flesh is white and has a rather sickly smell. Deadly poisonous. Urgent medical assistance should be sought if eaten. This species is called the Death Cap for good reason.
Range and habitat: A widespread species which is quite common in England. Grows especially under Oak but also several other deciduous trees such as Beech.
Similar species: The sickly green colour, white gills, ring and sac-like volva should make this species unmistakable.

FALSE DEATH CAP

Amanita citrina

Cap diameter: 4-8cm
Height: up to 7cm
Characteristics: The cap colour is generally off-white or lemon yellow and this darkens towards the centre. The cap is generally covered in large, fawn or brownish patches of tissue which are the remains of the veil. The gills are white and free. Spores white. Both the stem and the ring are white. The base of the stem is large and bulbous and is enclosed in the remains of the volva. This forms a shallow groove around the top of the bulb. The flesh is white and has a distinct smell of raw potato. Considered inedible on account of its unpleasant smell and taste.
Range and habitat: Much more common and widespread than its deadly relative. Found mainly under deciduous trees (especially Beech) but also under conifers as well.
Similar species: Distinguished from the true Death Cap by the flecks on the cap and the lack of sac-like volva.

DESTROYING ANGEL

Amanita virosa

Cap diameter: 5-10cm

Height: up to 12cm

Characteristics: The general appearance is of a pure white fungus, hence its angelic English name. The cap is conical in its early stages and when mature sometimes has a slight umbo. The gills are white, free and crowded. Spores white. The slender stem is often curved and is generally covered in woolly, fibrous hairs. The ring is white and is often rather tattered (sometimes appearing absent altogether). The bulbous base to the stem is enclosed in a sac-like volva. The flesh is white and has and unpleasant, sickly smell. Deadly poisonous and should be completely avoided.

Range and habitat: Rather uncommon. Grows mainly in deciduous woodland such as Oak, Beech and Birch but also in mixed woodland.

Similar species: Distinguished from other white *Amanita* fungi by application of Potassium hydroxide to the flesh which turns it yellow.

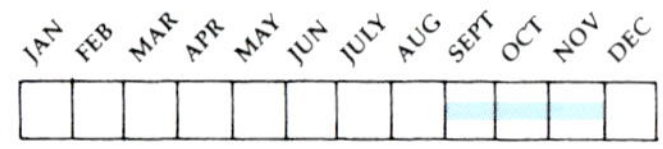

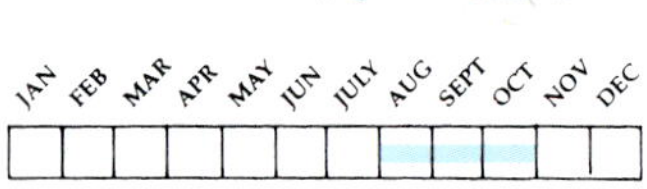

THE BLUSHER

Amanita rubescens

Cap diameter: 6-15cm

Height: up to 15cm

Characteristics: The domed cap becomes flattened as it expands. In young specimens, the mealy scales (remains of the veil) are densely packed. In mature Blushers, these become patchy and often rust-coloured in appearance. The cap colour is reddish brown or fawn and the gills are free and white, turning reddish where eaten by insects and slugs. Spores white. The ring is striated with fine lines and the stem is covered in reddish scales, becoming denser towards the base. The bulbous base lacks a volva. The English name derives from the fact that the white flesh gradually reddens when cut, bruised or exposed to air. Poisonous raw but considered edible and good by some when cooked. Best avoided.

Range and habitat: An extremely common fungus being found throughout Britain in most kinds of woodland.

Similar species: *A. excelsa* is similar but its flesh does not 'blush' when cut.

CAESAR'S MUSHROOM

Amanita caesarea

Cap diameter: 5-15cm
Height: up to 14cm
Characteristics: An attractive species whose appearance is matched by its delicious taste. The cap is bright orange or yellow and sometimes slightly sticky. Occasionally flecked with small fragments of the veil. The gills are yellow, free and crowded. Spores whitish. The stem is yellow. It bears a ring and has a bulbous base enclosed in a whitish volva. The flesh is yellow and has a pleasant smell. Edible and delicious. The English name reflects the esteem in which this fungus was held by the Romans. It is still a popular fungus in kitchens throughout mainland Europe.
Range and habitat: Restricted to the warmer parts of Europe and sadly not found in Britain. It is found in deciduous woodlands and especially under Oaks.
Similar species: Unlikely to be confused with any other species of *Amanita* because of the orange cap and yellow gills.

TAWNY GRISETTE

Amanita fulva

Cap diameter: 5-10cm
Height: up to 13cm
Characteristics: The cap is a rich tawny or orange-brown colour and almost flat in mature specimens. The colour darkens towards the centre where there is a slight umbo and the edge of the cap is striate. The gills are white and free. Spores white. The stem is slender and whitish and it lacks a ring; the bulbous base sits in a sac-like volva. Considered edible but perhaps best avoided.
Range and habitat: Widespread and occurs in deciduous woodland, often under Oak and Birch on acid soils. The Tawny Grisette can be extremely common in suitable habitats.
Similar species: The Grisette, *A. vaginata,* is very similar in size and shape to the Tawny Grisette. The most distinctive distinguishing feature is the cap colour, which is usually a grey-brown. The cap is 5-10cm in diameter and it stands up to 18cm tall. The Grisette is found in deciduous woodland, often under Beech.

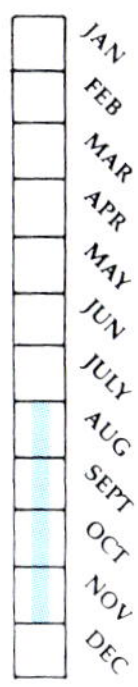

AMANITA CROCEA
Cap diameter: 5-10cm
Height: up to 15cm
Characteristics: A striking species of *Amanita* in which both the cap and the stem are bright orange-yellow. The surface of the cap is smooth and lacks any remains of the veil. In young specimens, the cap is domed. However, as they age they flatten out. The cap eventually rolls up at the margins and has a small umbo. The stem, which lacks a ring, is covered in rows of felty, orange hairs or scales giving it a woolly appearance. The base of the stem is not bulbous and is enclosed in a sac-like volva which is paler than the rest of the fungus. The gills are off-white as is the flesh which has a pleasant smell. Edible but not worth considering. Spores white.
Range and habitat: Widespread but locally distributed and nowhere common. Most frequently encountered in the north. The preferred habitat is under Birch but it will also grow in mixed, deciduous woodlands.
Similar species: Difficult to confuse with any other species.

AMANITA INAURATA
(= Amanitopsis strangulata)
Cap diameter: 5-15cm
Height: up to 14cm
Characteristics: A robust species in which the cap is rounded in the early stages but which expands to a flattened dome as it matures. The cap colour is greyish-brown, or sometimes richer brown, becoming paler and striated towards the edges. Large, dirty-grey, flaky patches (remains of the veil) usually remain attached to the surface. Both the gills and the flesh are white. Spores white. The stout stem lacks a ring and is flecked with bands of woolly tissue. The base of the stem is not noticeably bulbous and the volva is fragile and often disintegrates. Edibility uncertain. May be poisonous and so best avoided.
Range and habitat: A rather local and uncommon species which grows in mixed woodlands.
Similar species: Rather similar to the Grisette, *A. vaginata*, and the Tawny Grisette, *A. fulva*. However, these species generally lack the flaky patches on the cap surface.

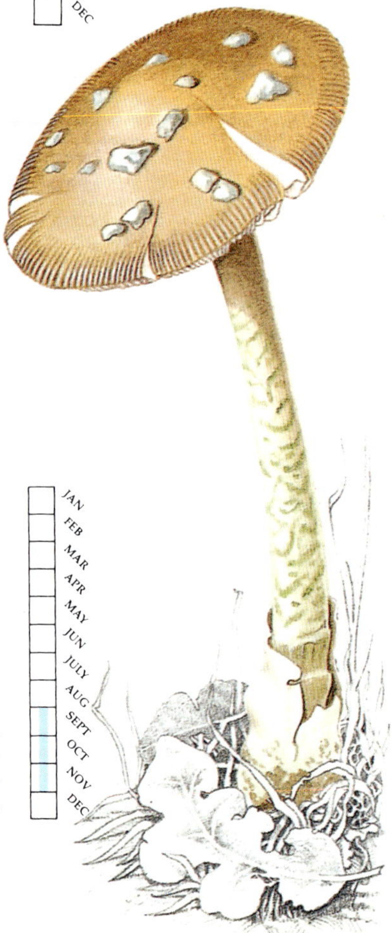

AMANITA EXCELSA

(= *Amanita spissa*)

Cap diameter: 5-12cm
Height: up to 12cm
Characteristics: The cap is a greyish-brown colour. At first this is covered by large patches of greyish-white tissue (the remains of the veil). In common with many species of *Amanita* these gradually get washed off by rain. Both the gills and the flesh are white. Spores white. The striated ring is whitish and often paler than the stout stem. The base of the stem is bulbous. The volva is difficult to distinguish and sometimes appears absent. Considered edible by some but best avoided because of confusion with poisonous species of *Amanita*.
Range and habitat: This fungus can be abundant along woodland rides and among leaf litter. Occurs in both deciduous and coniferous woodland.
Similar species: The Panther Cap, *A. pantherina*, has white (not grey) flecks on the surface of the cap and a rim around the base of the stem. The Blusher, *A. rubescens*, has flesh which reddens when cut or bruised.

PINK-SPORED GRISETTE

Volvariella speciosa

Cap diameter: 5-10cm
Height: up to 10cm
Characteristics: The cap becomes flattened in mature specimens. Colour usually pale grey-green but sometimes pale muddy brown, darkening towards the centre. The cap surface is extremely sticky, especially in damp weather. The gills are broad, closely packed and dull orange or pink in colour in mature specimens; the spores are also pink. The stem is whitish and tapers from the base which is slightly bulbous and enclosed in a fragile white volva. This disintegrates easily when picked. The flesh is white. Considered edible by some but probably best avoided.
Range and habitat: Widespread and often common in Britain and northern Europe. Grows on rotting straw bales, compost heaps, manured ground.
Similar species: *V. bombycina* has a paler cap covered in downy hairs. Cap is domed in maturity. Spores pink. *V. murinella* is smaller with a white cap.

VOLVARIELLA BOMBYCINA

Cap diameter: 10-20cm
Height: up to 18cm
Characteristics: The cap is large and remains domed even in mature specimens. The ground colour is pale yellow-brown, sometimes almost white. The surface is covered in fine, silky hairs which often stain with age. This gives the cap an attractive, felty appearance. The gills are broad and crowded. In young specimens they are white but soon become stained pink as the spores mature. The stem is white, lacks a ring and is often curved due to the growing angle. The bulbous base is enclosed in a white, sac-like volva. The flesh is white. Edible and delicious. Spores pink.
Range and habitat: Grows in small groups on rotting stumps of Elm and sometimes on dying trees. Occasionally on other deciduous species. A widespread species but not at all common. Most frequently encountered in southern England.
Similar species: Somewhat similar in appearance to *Pluteus cervinus* which lacks a volva and a felty cap.

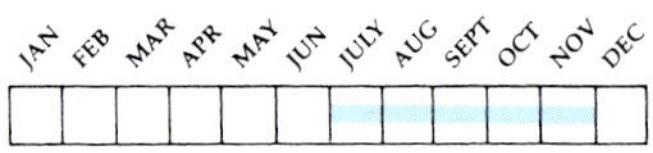

FAWN PLUTEUS

Pluteus cervinus
Cap diameter: 5-12cm
Height: up to 12cm
Characteristics: The cap forms an attractive, broad dome which develops a slight umbo in mature specimens. The colour is generally muddy brown to orange-brown and the surface is covered in radiating streaks. The gills are broad and whitish, turning pink as the spores mature. The stem is whitish and streaked with fine, dark lines. The base is slightly bulbous and lacks a volva. The flesh is white. Edible but not worth considering.
Range and habitat: Common and widespread throughout Britain. Although most frequent in the autumn, this species may be found at almost any time of year. It grows on rotting stumps of deciduous trees and sometimes even on the sawdust produced by woodland management.
Similar species: *P. salicinus* is a smaller species which has a blue-grey cap and a white stem. It also grows on rotting, deciduous timber and especially on Willow.

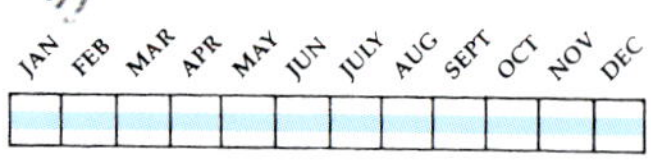

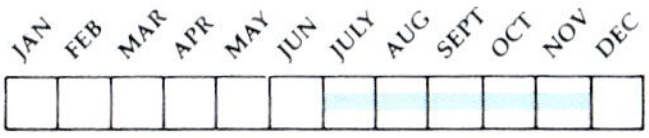

PARASOL MUSHROOM

Macrolepiota procera

Cap diameter: 10-25cm
Height: up to 30cm
Characteristics: A very distinctive fungus. The cap is a creamy-white to nut-brown colour. At first the cap is shaped like an organ-stop but it later expands to the size and shape of a plate with a distinct umbo. The surface is covered in dark, flaky scales. The gills are free, white and closely packed. Spores white. The stem, which bears a thick, double ring, is whitish covered in patches of darker scales. The flesh is white. Edible and delicious. Apart from the Field Mushroom, this is one of our most highly prized fungi.
Range and habitat: Found in open, grassy areas throughout Britain and northern Europe. Often grows in sizeable groups and fields, woodland rides and even wide roadside verges are favoured.
Similar species: The Parasol Mushroom is almost unmistakable but smaller species of *Macrolepiota* may resemble early stages of *M. lepiota*.

SHAGGY PARASOL

Macrolepiota rhacodes

Cap diameter: 7-14cm
Height: up to 16cm
Characteristics: The cap is egg-shaped in young specimens but expands with age to form a flattened dome. There is an indistinct umbo. The cap colour is off-white to pale brown and the surface is covered in patches of darker brown scales. These are often shaggy and tattered hence the English name of the fungus. The gills are broad and white and the flesh is also white. Spores white. The biscuit-brown stem is robust, bears a tough, double ring and lacks any dark markings or patches. Edible and delicious. Another highly prized species but one which may upset some people.
Range and habitat: Common and widespread. Found along grassy woodland rides, fields at woodland edges and even on garden lawns.
Similar species: *M. procera* is larger and has a marked stem. *L. mastoidea* is a paler, more delicate species found along woodland rides. *L. friesii* has an orange-brown cap with darker, scales.

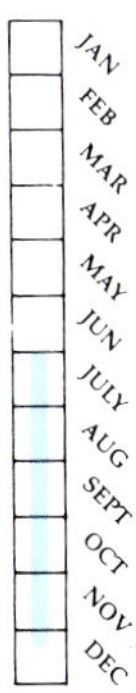

STINKING PARASOL

Lepiota cristata

Cap diameter: 2-6cm
Height: up to 4cm
Characteristics: The attractively marked cap has a background colour of off-white. The central umbo is a rich orange-brown colour. Scattered patches of similarly coloured scales, arranged in concentric rings, spread from the centre. The shape of the cap is usually irregular. The gills are white, free and closely packed. The spores are white. The stem is slender, pale-brown and carries a delicate ring. The base of the stem is not markedly swollen. Distasteful and unpleasant, both to smell and to taste, and may be poisonous. Should definitely be avoided. When bruised, the flesh has a distinct odour of tar.
Range and habitat: Grows along woodland paths, under hedges and even in gardens. Widespread and sometimes common in good years.
Similar species: *L. brunneoincarnata* is stouter and darker, smells of fruit, often has a rosy flush and is much less frequent.

SAFFRON PARASOL

Cystoderma amianthinum

Cap diameter: 2-5 cm
Height: up to 5 cm
Characteristics: An elegant species which is a distinct saffron-yellow or yellow-brown in colour. The cap is at first domed but flattens out as it expands. It may eventually have a dimpled centre with an umbo and margins which are irregularly wrinkled. The surface has a characteristic mealy or granular texture. The gills are white and crowded and the flesh is off-white. Spores white. The stem is the same background colour as the cap but below the ring is stippled with darker scales towards the clubbed base. Considered edible but best avoided.
Range and habitat: Widespread and sometimes locally common. Prefers heathland scrub and coniferous woodland.
Similar species: *C. carcharias* is a more northern species with similar habitat preferences to the Saffron Parasol. The colour, however, is off-white or flesh.

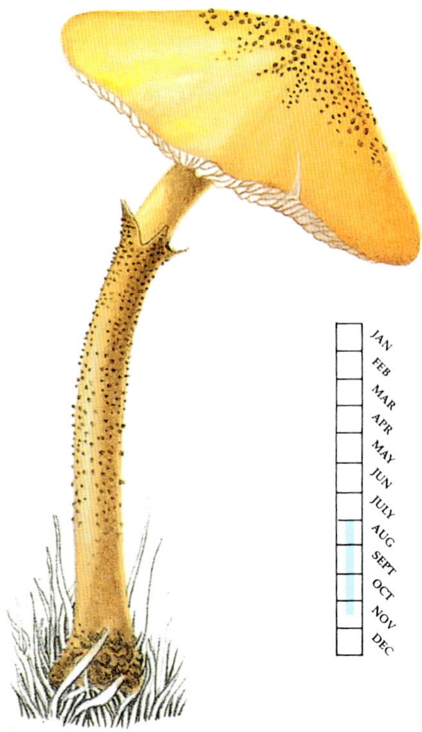

FIELD MUSHROOM

Agaricus campestris

Cap diameter: 5-10cm

Height: up to 10cm

Characteristics: The well-known 'wild' mushroom. The cap is, at first, button-shaped but soon expands to a wide dome. The colour is off-white or pale brown and the surface may be smooth or have large, flaky strips of tissue. The gills are pink in young specimens (button mushrooms), turning brown as the cap expands. Spores brown. The flesh is white but discolours slightly on exposure to air. It has the characteristic mushroom smell. The white stem has a ring which is often lost with age. Edible and delicious. This is one of the finest of our wild fungi and is widely sought.

Range and habitat: Common and widespread in suitable areas. Prefers pastures and meadows, often where cattle or horses have been grazing.

Similar species: *A. bisporus* (=*A. brunnescens*) has a slender stem with an obvious ring. Possibly the ancestor of the cultivated mushroom.

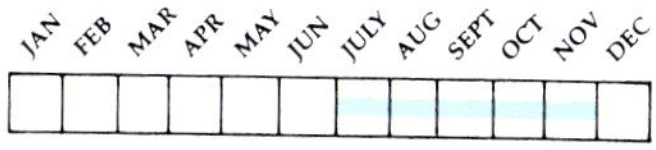

HORSE MUSHROOM

Agaricus arvensis

Cap diameter: 10-20cm

Height: up to 10cm

Characteristics: A stout species whose cap is cream or off-white in colour. May become marked or stained yellow as it ages. The initial shape of cap is globular but it expands to form a shallow dome. The gills are free and change from white to pale-pink to brown as the fungus matures. Spores dark brown. The off-white stem is robust and sometimes hollow. It has a ring comprising 2 membranes, the lower one generally splitting and appearing toothed. The flesh is white. Edible and delicious. Much prized on account of its wonderful flavour and impressive size.

Range and habitat: A widespread and common species. Found in grassy meadows, pastures and lawns.

Similar species: *A. bitorquis* often occurs on roadside verges and has a flattened cap and narrow gills. *A. augustus* is large and has a dull orange cap covered in scales and a woolly stem.

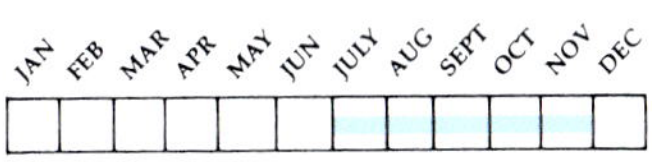

WOOD MUSHROOM

Agaricus sylvicola

Cap diameter: 6-12cm

Height: up to 9cm

Characteristics: The cap is a creamy-white colour which gradually becomes yellow with age and damage. The cap is often rather flat in older specimens and may split around the margins. The gills are free and change from off-white to rich brown with age. Spores dark brown. The stem is the same colour as the cap and has a prominent ring comprising two separate membranes. The lower membrane often has a serrated edge. The base of the stem is somewhat bulbous. Edible and delicious. The flesh is white and smells of aniseed when fresh.

Range and habitat: As the English name suggests, this species favours woodlands, both deciduous and coniferous, and grows amongst leaf litter. It is common and widespread in Britain and northern Europe.

Similar species: *A. macrosporus* is a larger species more common in the north. It prefers upland pastures and has a faint smell of aniseed.

YELLOW STAINER

Agaricus xanthodermus

Cap diameter: 7-15cm

Height: up to 15cm

Characteristics: The domed cap becomes flattened with age. The general colour of the cap is off-white or pale-brown but when cut or bruised it stains a bright, sulphur yellow. This is particularly noticeable around the margin, which becomes split with age. The flesh is white but stains bright yellow at the base of the stem and underneath the cap. The gills darken with age to become a dark brown colour. Spores dark brown. The white stem supports a single, thickened ring and has a bulbous base often stained yellow. Poisonous. Although some people are not affected, avoid this readily identifiable fungus.

Range and habitat: A common and widespread species in woodland borders, fields, parks and lawns.

Similar species: This is the only species of *Agaricus* that stains bright yellow both in the stem and the cap. Should cause little confusion.

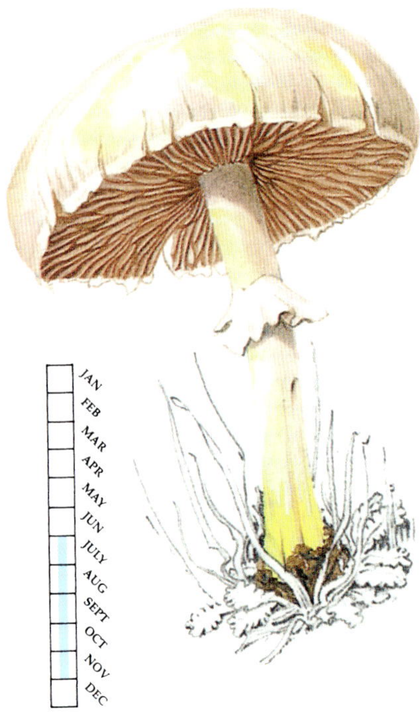

SCALY WOOD MUSHROOM

Agaricus sylvaticus

Cap diameter: 5-10cm
Height: up to 10cm
Characteristics: A rather distinctive species. The domed cap is covered in dark brown scaly fibres which mask the cap cuticle's background colour of pale brown. In some specimens, the scales give the appearance of short, matted hair. The gills are densely packed and change from pale pink to dark brown with age. Spores brown. The stem is off-white or pale brown in colour. It is often rather scaly below the rather shaggy ring. The stem base is bulbous. Edible and good. Flesh white turning red when cut or bruised.
Range and habitat: A common and widespread species in Britain and northern Europe. Found mostly in coniferous woodland but also occasionally in mixed woodland.
Similar species: *A. langei* also has white flesh which turns red when damaged. The larger size and stouter stem, lacking a bulbous base, help distinguish this species from *A. sylvatica.*

JAN	FEB	MAR	APR	MAY	JUN	JULY	AUG	SEPT	OCT	NOV	DEC

COMMON INK CAP

Coprinus atramentarius
Cap height: 4-8cm
Height: up to 16cm
Characteristics: In young stages, the cap is thimble-shaped but the margins spread and split as it ages. The colour is grey-brown, darkening towards the centre which is often slightly scaly. The overall colour tends to darken with age. The gills are free and crowded. Initially they are pale but they soon turn black and liquify as the spores are liberated. Spores black. The stem is smooth and white with a bulbous, noticeably ridged, base. Edible but best avoided. The fungus is a source of antabuse, a drug used to cure alcoholics. If consumed with alcohol, it causes nausea and heart palpitations. The liquifying caps were used as ink.
Range and habitat: Common and widespread. It often grows in large clusters on lawns, roadside verges or woodland rides, generally in close proximity to buried, rotting wood.
Similar species: *C. micaceus* has a nut-brown cap and grows in clumps.

SHAGGY INK CAP; LAWYER'S WIG

Coprinus comatus
Cap height: 5-15cm
Height: up to 30cm
Characteristics: A familiar species whose English name aptly describes its appearance. The cap is at first cylindrical in shape, white in colour and covered in shaggy fibres and scales. The centre of the cap is rather brown. As it ages, the cap expands to become bell-shaped and darkens, particularly around the margins when the gills start to liquify. The gills are initially pale pink but soon blacken. Spores blackish. The stem is slender and white and possesses a delicate ring. Edible and good. Only worth considering before the gills have begun to liquify.
Range and habitat: Widespread and common in Britain and northern Europe. Occurs in almost any grassy place including lawns and roadsides.
Similar species: The Magpie Cap, *C. picaceus*, is smaller and less common. The mature cap is dark with pale patches of veil tissue (see page 17).

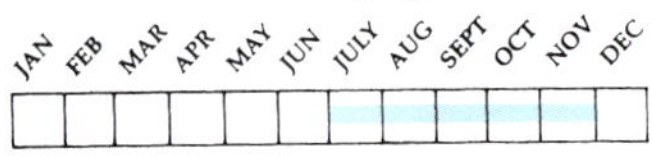

GLISTENING INK CAP

Coprinus micaceus
Cap height: 3-5cm
Height: up to 10cm
Characteristics: A conical or thimble-shaped cap which is nut-brown in colour, darkening towards the middle. The centre of the cap is peppered with light brown granules and the cap surface is grooved from the centre. The margins often split with age. The gills are white at first but soon turn black as the fungus matures. The gills and cap do not liquify to the same extent as many other species of *Coprinus*. Spores dark. The slender stem is white and smooth. Edible but not outstanding. The flesh is pale at first but darkens as the fungus ages.
Range and habitat: Common and widespread throughout Britain and northern Europe. Often found in extremely large clumps on the rotting stumps and logs of deciduous trees.
Similar species: *C. domesticus* is a more robust species and generally paler in appearance. It is much less common than *C. micaceus*.

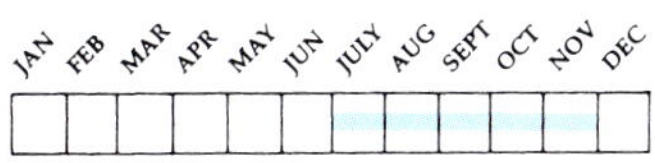

FAIRIES' BONNETS

Coprinus disseminatus
Cap height: 0.5-2cm
Height: 2-4cm
Characteristics: The cap is thimble-shaped at first but often expands with age. The cap structure is membranous, rather delicate and the surface is strongly grooved. Cap colour is grey-brown or buff, darkening towards the centre. The gills are white at first but turns black as the fungus matures. The gills do not liquify. The spores are a rich brown colour. The stem, which is often hard to detect in dense groups, is thin and whitish and covered in fine, white down near the base. Not poisonous but too small to be worth considering.
Range and habitat: Extremely common in southern Britain but less so further north. Grows in large, densely-packed clumps (frequently 50 or more) on the rotting stumps of deciduous trees, even when these are slightly buried in the soil.
Similar species: The dense colonies and size and shape of this species should make it readily identifiable.

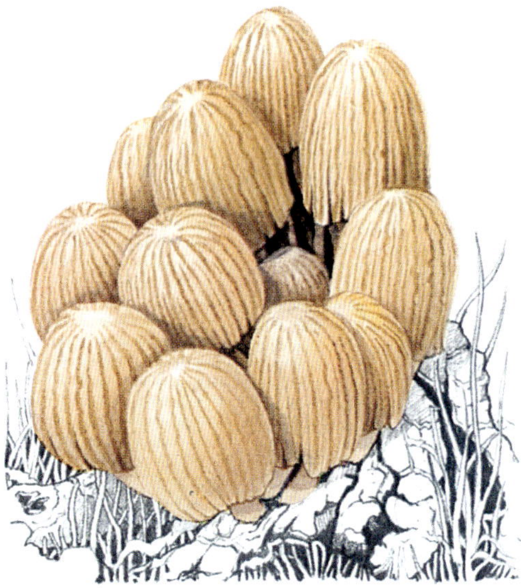

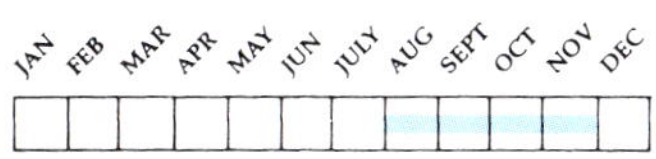

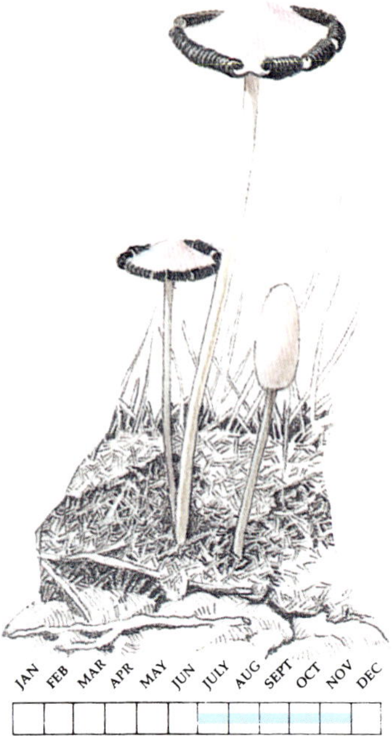

COPRINUS NIVEUS
Cap diameter: 2-4cm
Height: 4-8cm
Characteristics: A very distinctive fungus with a characteristic chalky appearance. The cap is at first conical or cylindrical but later expands and flattens. The edges roll back and blacken with age and the margin often splits. The cap surface is covered in a mealy powder like chalk and the colour is white. The gills are white at first but soon blacken as the fungus matures and they begin to liquify. Spores black. The stem is slender and white and lacks a ring. It shares the same mealy texture as the cap. Edibility unknown. Should therefore be avoided.
Range and habitat: A common and widespread species in suitable habitats. Only grows on horse or cow dung.
Similar species: The slender, white appearance of *C. niveus* and its habitat are good identification features. *C. lagopus* is also a pale, slender species. It grows among woodland leaf litter and has a hairy cap.

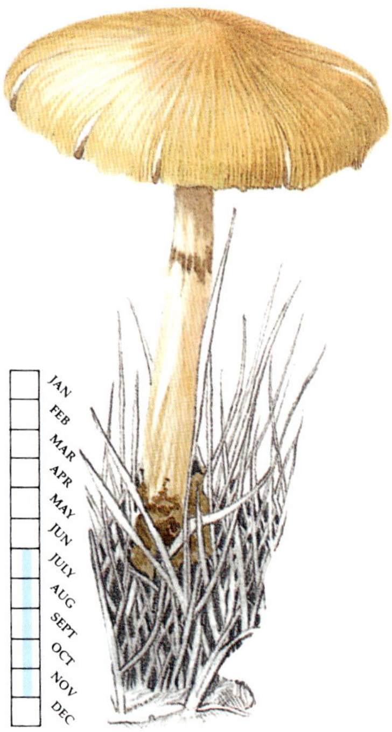

WEEPING WIDOW
Lacrymaria velutina
Cap diameter: 3-10cm
Height: 4-9cm
Characteristics: The cap is covered with woolly fibres and is broadly flattened when mature. The cap colour is buff or tan becoming brighter towards the umbonate centre. The colour darkens, particularly around the margin, as the fungus ages. The gills are crowded and dark with white edges. The English names derives from the gill margins which sometimes 'weep' in fresh specimens. Spores black. The stem is buffish-white and is covered in woolly fibres. Edible but an acquired taste and not to many people's liking. The flesh is brown and has a bitter taste.
Range and habitat: Common and widespread, sometimes seen in small clumps. Grows beside woodland rides and roadsides, usually in association with buried, rotting wood.
Similar species: *L. pyrotricha* has a brighter orange cap and grows in association with rotting coniferous stumps and logs.

CRUMBLE TUFT

Psathyrella candolleana

Cap diameter: 4-7cm
Height: up to 8cm
Characteristics: The cap is thimble-shaped when young but eventually broadens to a flattened cone shape. There is sometimes a slight umbo. The colour is generally pale-brown but can be almost white or darker buff in some specimens. This background lightens as the fungus dries. The cap margins are sometimes slightly striated. They often split and usually bear the tattered remains of the veil. The gills are greyish lilac, darkening as they age. Spores dark brown. The hollow stem is white, fragile and rather brittle. Edibility unknown. Not worth considering for eating.
Range and habitat: A common and widespread species found in gardens, woodland borders and beside overgrown hedgerows. It is usually associated with rotting tree stumps and often grows in clumps.
Similar species: *P. squamosa* is a smaller species with a woolly cap and is found in association with Beech.

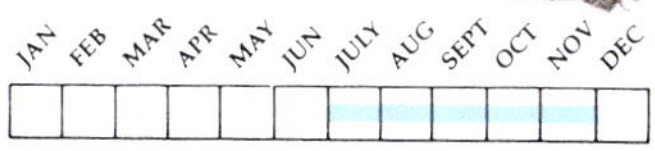

PSATHYRELLA HYDROPHILA

Cap diameter: 2-5cm
Height: up to 10cm
Characteristics: The cap has a rounded, convex shape sometimes becoming rather flattened. The general colour is tan to rich brown but as the fungus dries the centre of the cap fades to an orange-buff colour. The margins often bear the tattered remains of the veil. The crowded gills are at first buff in colour but soon turn a rich brown colour. Spores dark brown. The whitish stem sometimes bears the remains of the ring and becomes increasing buff coloured towards the base. Considered edible by some but has a bitter taste and so best avoided.
Range and habitat: A common and widespread species in Britain and northern Europe. Grows in tufts, often large and dense, in rich soils under deciduous woodland. Sometimes associated with tree stumps.
Similar species: *P. multipedata* is less frequently encountered. The cap is pale grey-brown in colour, fading towards the centre. Similar habitat.

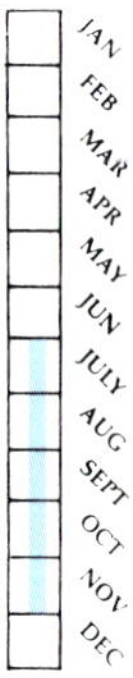

BROWN HAY CAP

Panaeolina foenisecii
Cap diameter: 1-3cm
Height: 4-6cm
Characteristics: The cap is bell-shaped, sometimes becoming rather flattened with age. The general coloration is rich brown but, as with *Psathyrella hydrophila*, the cap dries paler towards the centre. This often gives the cap a rather zoned appearance, dark around the margin with a buff-tan middle area and the very centre sometimes orange-tan. The gills start pale brown but become mottled darker with age. Spores black. The slender stem is pale brown. Not an edible species.
Range and habitat: A common and widespread species in Britain and northern Europe, often occurring on garden lawns. Also found in parks, fields and beside roadsides, in fact almost anywhere with short grass.
Similar species: *Panaeolus ater* is a similar size but the stem is stouter and dark brown, similar in colour to the cap. Lawns and other areas of short grass.

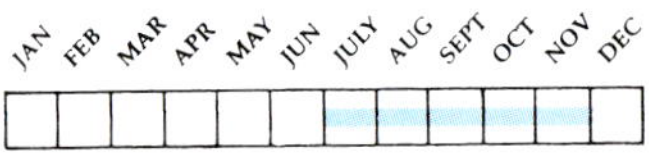

DUNG FUNGUS

Panaeolus semiovatus
Cap diameter: 3-6cm
Height: up to 10cm
Characteristics: The cap has a distinctive rounded bell shape which resembles half an egg shell. The colour is usually off-white or buff, becoming tanned towards the centre. In wet weather the smooth cap becomes rather sticky. The remains of the veil often remain attached to the margin. The broad gills are pale at first but soon blacken. Spores black. The stem, which tapers slightly, bears the remains of the ring. It is often curved at the base due to the growing position. Not edible.
Range and habitat: Grows only on horse and cattle dung. Therefore has a rather patchy distribution but nevertheless common where it does occur. More frequent as you move north in Britain.
Similar species: The size, colour and habitat of this species make it easy to identify. *P. speciosus* is smaller, darker and with a shallow, distinctly flattened cap.

PANAEOLUS SPHINCTRINUS

Cap diameter: 2-4cm
Height: up to 10cm
Characteristics: The conical cap is smooth and usually dark grey. Occasionally, pale buff specimens are found. As the cap dries, the colour lightens. The margin is broadly serrated or toothed due to the presence of veil remains. These, however, are sometimes lost in older specimens. The gills are grey at first but soon become mottled and blackened. They are edged with white. Spores black. The slender stem is smooth and pale grey, darkening towards the base. Edibility uncertain and so best avoided.
Range and habitat: A common and widespread species in Britain and northern Europe in suitable habitats. It is found in meadows and pastures, usually on horse or cow dung.
Similar species: *P. campanulatus* also grows in association with horse dung but is less common. The cap is pale brown rather than grey and it has a more domed appearance than *P. sphinctrinus.*

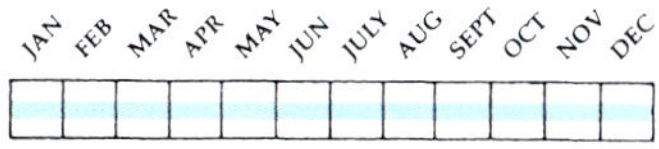

SULPHUR TUFT

Hypholoma fasciculare
Cap diameter: 4-8cm
Height: 5-10cm
Characteristics: A very familiar and well-named species. The domed cap is bright sulphur yellow in colour. The centre is usually deeper orange and there is sometimes a slight umbo. The caps are often closely packed together. The gills are sulphur yellow at first but become olive-green and soon darken with age. Spores purple brown. The tough, fibrous stem is often twisted or curved. The colour is yellow but it darkens towards the base and shows the remains of the ring. Not edible. The flesh is sulphur yellow and the taste is bitter.
Range and habitat: An extremely common and widespread species. Grows in large clumps or tufts (sometimes more than 100 together) on the rotting stumps or logs of deciduous trees.
Similar species: *H. marginatum* is similar but has a tan not sulphur cap, pale stems cloaked in white hairs and grows under conifers. Not as common and widespread as *H. fasciculare.*

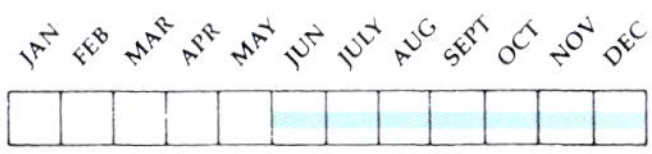

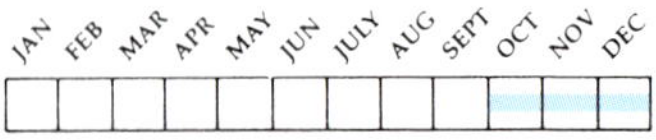

BRICK CAPS

Hypholoma sublateritium

Cap diameter: 5-10cm

Height: up to 15cm

Characteristics: As the English name suggests, the broad, flattened dome-shaped cap is generally brick-red to deep brown in colour. Towards the margin, the colour becomes paler sometimes forming a distinct band. The gills are pale yellow at first but darken to grey-brown with age. Spores lilac brown. The tough stem is pale brown becoming darker towards the base. Not edible. The yellow-brown flesh has a bitter taste.

Range and habitat: Widespread and locally common in Britain and northern Europe. Grows on rotting stumps of deciduous trees. Often forms large clumps with the caps closely packed together.

Similar species: The Conifer Sulphur Tuft, *H. capnoides*, has an appearance midway between *H. fasciculare* and *H. sublateritium*. The cap is tan coloured fading to a buff margin. It grows in clumps on conifer stumps. Not edible.

LIBERTY CAP

Psilocybe semilanceata

Cap diameter: 1-2cm

Height: up to 6cm

Characteristics: This is the so-called 'magic mushroom'. The cap has a distinctive and characteristic shape. It forms a rounded cone with a pronounced and pointed umbo. The cap colour is pale-buff or olive-tan and the umbo is sometimes a deeper brown. In maturity, the margin sometimes acquires a blue-green tint. The cap surface is smooth and slightly sticky. The gills are buff at first but soon darken to a purple-brown colour. Spores brown-purple. The thin stem is often wavy and is an olive-tan colour. Not to be eaten since this fungus has marked hallucinogenic properties.

Range and habitat: A widespread and locally common species in Britain and northern Europe. It is found on areas of short grass such as lawn, parks, fields and even roadside verges.

Similar species: Several other less frequent species of *Psilocybe* occur in the region but they lack the pointed umbo of *P. semilanceata*.

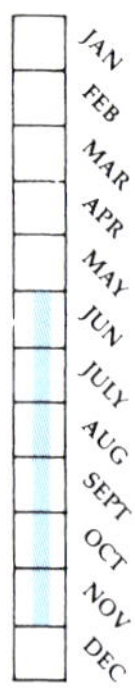

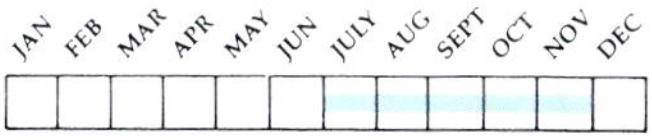

VERDIGRIS AGARIC

Stropharia aeruginosa
Cap diameter: 2-8cm
Height: up to 10cm
Characteristics: The cap is at first domed and rounded but expands to form a flattened cone with age. Cap surface is blue-green in colour (hence the English name), extremely slimy and covered in white scales when young. There is a slight umbo which is generally yellowish in colour. The gills are white at first but turn dark brown. The spores are purple-brown. The stout stem is bluish-white in colour. It is covered in whitish, flaky scales below the dark ring. Above this, the stem is rather smooth. Not edible and may even be poisonous.
Range and habitat: A common and widespread species. It grows in woods, meadows, hedgerows and in mature gardens.
Similar species: *S. coronilla* has an orange-yellow cap and white stem and gills. The ring, which is also white, traps the dark spores and may appear brown. Common in grassy habitats, on lawns and in meadows.

DUNG ROUNDHEAD

Stropharia semiglobata
Cap diameter: 1-4cm
Height: up to 10cm
Characteristics: An aptly named species which both grows on dung and generally has a distinctive hemispherical cap. The cap colour is creamy-yellow or buff with the centre often stained darker. The cap surface is slightly sticky and the gills are purple-brown. Spores purple-brown. There is sometimes a slight umbo. The slender, sometimes sinuous, stem is pale yellow and bears the remains of a dark ring. Below the ring, the stem is sticky. Not edible.
Range and habitat: A common and widespread species in Britain and northern Europe. Grows actually on dung or in well-manured fields grazed by cows or horses.
Similar species: The rounded cap and habitat are distinctive in this species. *S. hornemannii* is a much larger species with a shorter, stouter stem and a sticky cap. It grows in deciduous woodland among fallen twigs or sawdust.

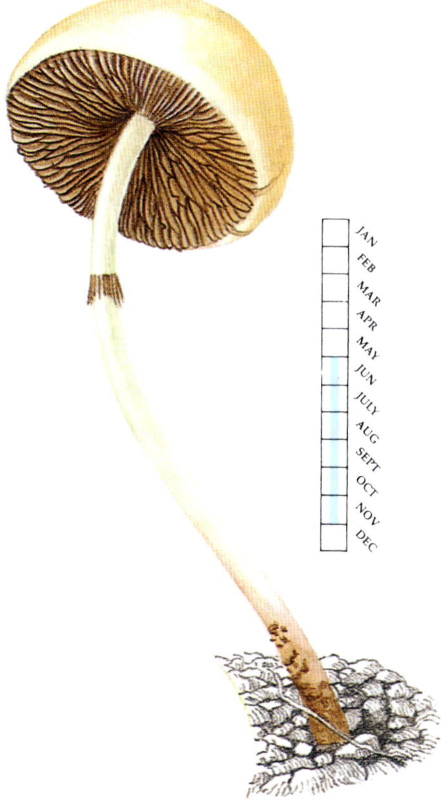

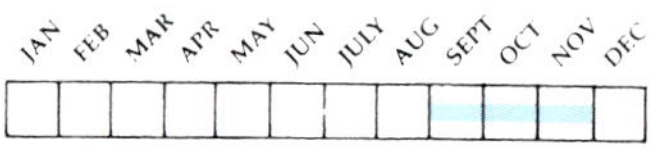

PHOLIOTA FLAMMANS

Cap diameter: 3-8cm

Height: up to 8cm

Characteristics: The cap is at first rounded but expands to become broadly flattened. It is dull yellow in colour but is covered in sulphur or lemon-yellow shaggy and recurved scales. The cap sometimes has a slight umbo and the margins are inrolled. The gills are yellow but darken with age. Spores red-brown. The yellow stem bears a ring of the same colour. The stem is cloaked in fibrous scales below the ring but rather smooth above. Not considered edible.

Range and habitat: A northern species which is at its most frequent in the Highlands of Scotland. Grows in clumps on rotting stumps and branches of conifers.

Similar species: *P. adiposa* is a similar species but has a duller yellow cap covered with darker fibrous scales. It grows in clumps on the decaying stumps and logs of deciduous trees, especially Beech, and is widespread but local throughout the region.

Pholiota adiposa

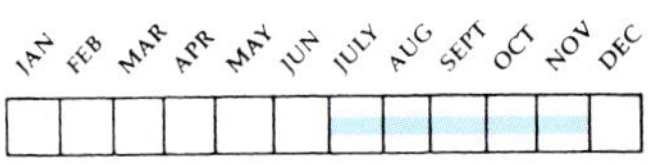

SHAGGY PHOLIOTA

Pholiota squarrosa

Cap diameter: 5-10cm

Height: up to 12cm

Characteristics: A distinctive species whose cap is domed at first but soon expands and becomes flattened. The colour is pale-brown or buff and the cap surface is covered in nut-brown, fibrous scales which are recurved. Always retains its shaggy appearance. There is generally a slight umbo and the cap margins are inrolled. The gills are yellow at first but soon darken. Spores rust brown. The stem is tough and often curved. Above the shaggy ring the stem is smooth but below, it is covered in brown, fibrous scales. Not edible. The flesh is yellow or off-white.

Range and habitat: A very common and widespread species which grows in clumps at the base of deciduous trees.

Similar species: *P. lubrica* is similar but has a slightly sticky red cap, much less shaggy in appearance. The stem has fewer, smaller scales. Grows at the base of deciduous trees, especially Beech.

COMMON WHITE INOCYBE

Inocybe geophylla

Cap diameter: 1-3cm

Height: up to 5cm

Characteristics: The cap is white and conical at first but it soon expands and develops a pronounced umbo. With age, the cap becomes stained and blotched with pale yellow. The surface is smooth and silky. The crowded gills are off-white at first but gradually darken with maturity. Spores brown. The stem is white and smooth. Poisonous. The flesh is white and has a rather unpleasant, sickly smell.

Range and habitat: A common and widespread species throughout Britain and northern Europe. Grows along forest rides and tracks in deciduous and mixed woodland.

Similar species: *I. geophylla* sometimes occurs as a lilac form, var. *lilacina*. Apart from the colour, it is similar in all other respects. *I. griseolilacina* has a muddy-brown cap with violet tints and a scaly surface. The flesh is lilac-grey. It grows in deciduous woodlands, often beside paths and rides.

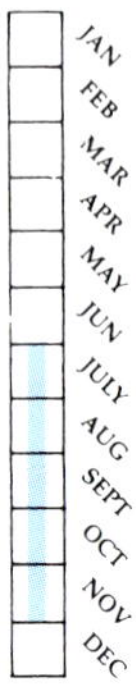

INOCYBE COOKEI
Cap diameter: 2-5cm
Height: up to 6cm
Characteristics: In its early stages, the cap is conical with a slightly inrolled margin. It soon expands and flattens and develops a pronounced umbo. The cap margin is sometimes faintly striated and readily splits with age. The cap is straw-coloured or pale brown. The gills are whitish but soon darken with age. Spores brown. The stem is off-white, darkening towards the base which is markedly bulbous with a ridged margin. Poisonous. The flesh is off-white to pale brown with a faint sweet smell.
Range and habitat: A common and widespread species in Britain and northern Europe. It grows among fallen leaves in deciduous and mixed woodlands.
Similar species: *I. napipes* has a grey-brown, conical cap with a distinct umbo and a pale stem. The base of the stem is bulbous but it lacks the ridged margin of *I. cookei*. Poisonous. Grows in mixed woodlands throughout the region.

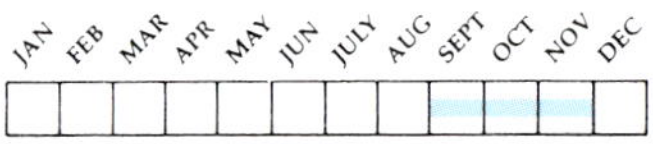

INOCYBE MACULATA
Cap diameter: 4-8cm
Height: up to 9cm
Characteristics: The cap is conical at first but later becomes flattened with a pronounced umbo. The margin frequently splits in mature specimens. The cap surface is radially streaked with dark brown, fibrous hairs and the downy, white remains of the veil sometimes adorn the umbo. The gills are off-white at first but darken. Spores brown. The stem is pale brown, sometimes with a streaked appearance. The base of the stem is not bulbous. Poisonous. The flesh is white with a strong smell.
Range and habitat: A common and widespread species. Grows in deciduous woodland especially under Beech.
Similar species: *I. fastigiata* is similar in size and shape. The cap is creamy-buff, readily splits at the margin and is radially streaked with thicker brown lines than *I. maculata*. The stem is white, sometimes stained buff under the cap. A common and widespread species, growing in deciduous woodland.

RED-STAINING INOCYBE

JAN FEB MAR APR MAY JUN JULY AUG SEPT OCT NOV DEC

Inocybe patouillardii

Cap diameter: 3-8cm

Height: up to 10cm

Characteristics: At first the cap is conical but it expands and flattens showing a prominent umbo. The margin splits deeply. The cap is creamy-white or flesh coloured and is radially streaked with reddish-brown fibres. The gills are pinkish-white becoming olive-brown. The spores are dull brown. The stem is whitish. All parts of the fungus bruise or stain red with age as the English name suggests. Poisonous. This is one of our most deadly species.

Range and habitat: This is widespread but rather local species. It grows in clearings and beside rides and paths in deciduous woodland, especially under Beech on chalk soils.

Similar species: *I. godeyi* is similar in size and shape. The general colour is tan, although the stem may be paler. The cap is streaked and does not split as readily as *I. patouillardii*. The stem has a bulbous base. Habitat similar to *I. patouillardii*.

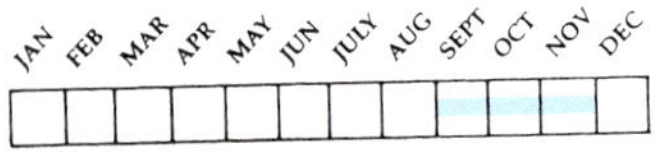

HEBELOMA MESOPHAEUM

Cap diameter: 2-5cm
Height: up to 8cm
Characteristics: The cap is domed at first but becomes flattened with age. The cap is generally reddish-brown and viscid near the centre with a paler margin covered with the fibrous remains of the veil. The margins of the cap often split with age. The gills are pale muddy-brown. Spores brown. The stem, which bears the fibrous remains of the ring, is pale brown and has a slightly fibrous texture. Sometimes paler above the ring remains than below. Edibility uncertain and so best avoided. The flesh is white, becoming stained brownish towards the base of the stem.
Range and habitat: A common and widespread species in Britain and northern Europe. Generally grows in association with Birch and so often found on wooded heaths.
Similar species: *H. sacchariolens* has a paler cap which is still darker towards the centre. There is a slight umbo. Less frequent in damp woodland throughout the region.

POISON PIE

Hebeloma crustuliniforme
Cap diameter: 5-10cm
Height: up to 8cm
Characteristics: The cap is broadly flattened and has a shallow umbo. The cap surface is smooth and slightly sticky in damp weather. The cap colour is buff or flesh around the margin, darkening to tan or reddish-brown towards the centre. The gills are pale brown and exude droplets of water in damp weather. Spores rich-brown. The robust stem is off-white in colour. Poisonous, as the English name suggests. The flesh is white and smells of radish.
Range and habitat: A common and widespread species in Britain and northern Europe. Grows in deciduous and mixed woodlands and overgrown hedgerows.
Similar species: *H. sinapizans* is considerably larger but also smells of radish. The domed, slightly sticky cap is generally a uniform tan colour and the stem is whitish. The brown gills exude water droplets. Rather uncommon. It grows in deciduous woodland.

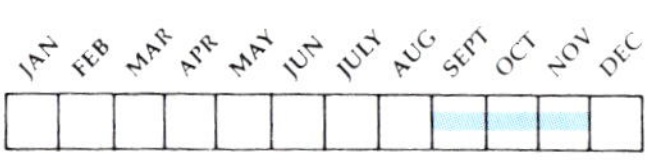

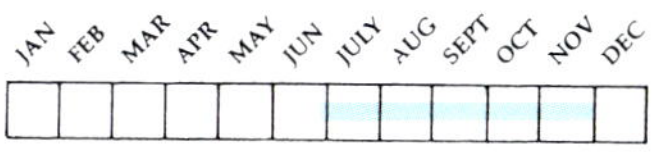

CORTINARIUS PSEUDOSALOR

Cap diameter: 5-9cm
Height: up to 10cm
Characteristics: The cap is conical at first. Later expands and becomes flattened with a shallow umbo and a wrinkled margin. The cap colour is buff to tan, reddening and darkening towards the centre. The surface is sticky and glutinous in damp weather but is shiny when dry. The gills are a muddy-brown colour later becoming rusty or tan. The spores are rust brown in colour. The whitish, tapering stem is sticky and often tinted violet. Possibly edible but best avoided.
Range and habitat: A common and widespread species in Britain and northern Europe. Grows in mixed and deciduous woodland, especially under Beech.
Similar species: *C. collinitus* has a yellow brown cap, becoming orange-tan towards the centre. The gills are pale becoming rust-brown. The stem is whitish but often tinted violet. Less frequent than *C. pseudosalor.* Generally found in mixed or coniferous woodland.

RED-BANDED CORTINARIUS

Cortinarius armillatus
Cap diameter: 6-12cm
Height: up to 15cm
Characteristics: The cap is domed or bell-shaped at first. Later expands and flattens sometimes with a slight umbo and an upturned margin. The cap colour is orange-tan or rusty-brown in older specimens, generally darker towards the centre. The gills are orange-tan becoming darker tan with age. Spores rust brown. The stem is pale-brown and bears the remains of the veil which form one to three reddish-brown bands. The base of the stem is distinctly bulbous. Edibility unknown and so best avoided.
Range and habitat: A widespread and locally common species throughout the region. Generally grows on acid soils in association with Birch and so found on heaths and mixed woodland.
Similar species: *C. bulliardii* is much less common. The cap and gills are a rusty-brown colour. The reddish-brown stem has a bulbous base and lacks red bands. Grows in deciduous woodland, especially under Beech.

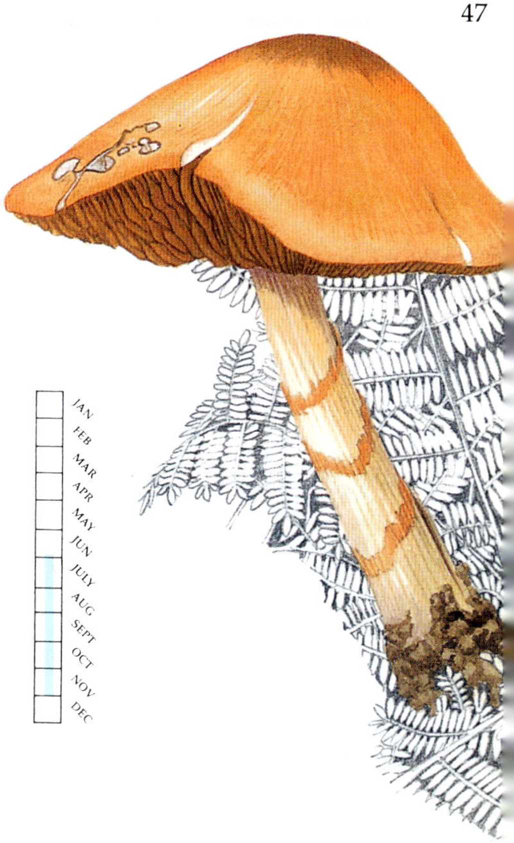

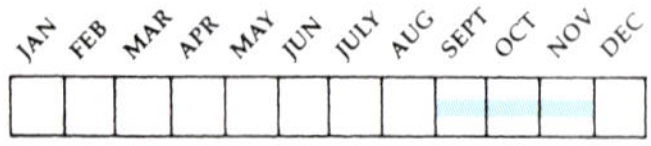

CORTINARIUS GLAUCOPUS

Cap diameter: 5-10cm
Height: up to 10cm
Characteristics: The cap is at first hemispherical but later becomes a flattened dome. Colour is orange-brown, often blotched paler and paler around the margins. Generally radially streaked with darker, wavy fibrous lines. The surface is slightly sticky. The gills are blue at first but become brown with age. Spores rust brown. The short, stout stem has a swollen base and the fibrous remains of the veil often remain attached to the expanding cap margin in young specimens. The stem often bears brown, fibrous lines. Edibility unknown and so best avoided. Flesh white, becoming blue at stem apex.
Range and habitat: A rather scarce species found mainly under conifers but sometimes in mixed woodland.
Similar species: *C. auroturbinatus* has a golden-yellow cap which is paler at the margin. The stem has a bulbous base and is marked with dark fibrous lines. Grows under Beech trees on chalk soils. Locally common.

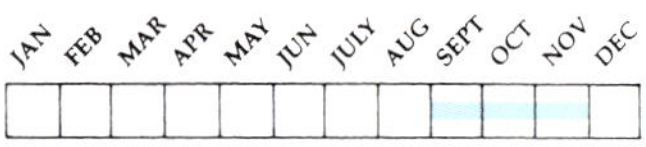

CORTINARIUS ALBOVIOLACEUS

Cap diameter: 4-8cm
Height: up to 10cm
Characteristics: A rather distinctive species. The cap is bell-shaped at first but expands to a flattened dome sometimes with a shallow umbo. The cap colour is bluish-white with lilac tints. Older specimens are sometimes almost white. The cap has a silky texture. The gills are bluish at first but become muddy-brown with age. Spores rust brown. The stem is the same colour as the cap with deep-violet fibrous lines. The stem tapers and the base is slightly swollen. Edibility unknown and so best avoided. The flesh is bluish-white.
Range and habitat: Widespread in Britain but nowhere particularly common. Grows in deciduous woodland, especially under Oak and Beech
Similar species: *C. traganus* is superficially similar but the colour of the cap and stem are generally deeper violet. The flesh is yellow. A distinctly northern species, most frequently found growing under conifers in Scotland.

CORTINARIUS PURPURASCENS
Cap diameter: 10-15cm
Height: up to 14cm
Characteristics: The cap is at first domed and hemispherical but later becomes flattened with a broad, shallow umbo. The cap colour is dirty-brown, sometimes tinged with violet. The cap is sticky and shiny in damp weather. The gills are violet at first becoming brown later. They bruise deep violet. Spores rust brown. The stem is dirty-brown with violet tints, often with dark streaks, and the base is bulbous. May be edible but not recommended and best avoided. The whitish flesh is tinted violet but stains deep-purple when bruised.
Range and habitat: A common and widespread species in Britain and northern Europe. Found in both coniferous and deciduous woodland.
Similar species: *C. sodagnitus* has a bluish cap 3-6cm in diameter which stains yellowish towards the centre. The whitish flesh does not discolour when bruised. Locally common under Beech trees on chalk soils.

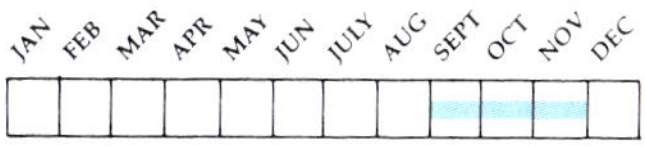

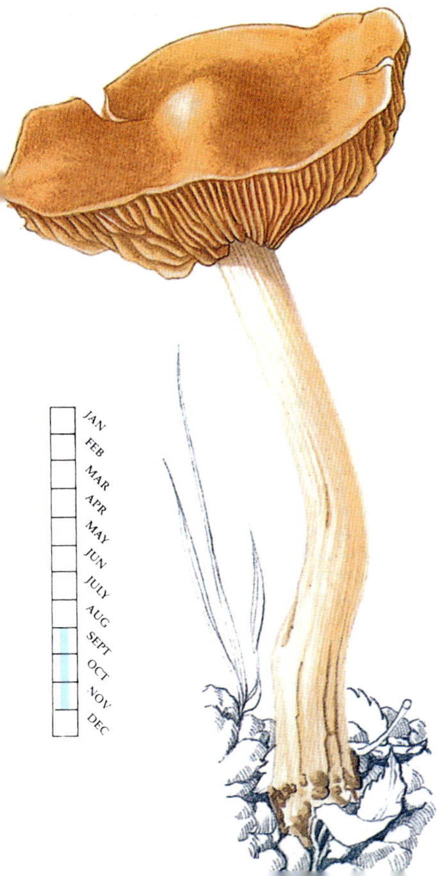

CORTINARIUS ANOMALUS
Cap diameter: 4-8cm
Height: up to 10cm
Characteristics: The cap is at first domed but later becomes broadly flattened with a shallow umbo. The colour is brownish violet at first but becomes tan-coloured with age. The surface is slightly shiny. The gills are violet at first but soon become tan or orange-brown. Spores rust brown. The stem is whitish, sometimes tinged with violet. The base is bulbous and the stem often bears dirty-brown rings. Edibility doubtful and so best avoided. The stem is white, staining towards the base.
Range and habitat: A common and widespread species in Britain and northern Europe. Grows under both deciduous and coniferous trees and often associated with Birch.
Similar species: *C. amoenolens* has a buff or pale-tan cap 6-12cm in diameter. The gills are violet then dark brown and the white stem has a bulbous, yellowish base. A rather local species. Grows under Beech trees on chalk soils.

GYMNOPILUS PENETRANS

Cap diameter: 3-6cm
Height: up to 7cm
Characteristics: The cap is conical at first but becomes flattened with age without a noticeable umbo. A wrinkled cap margin develops as the cap expands. The cap colour is a rich golden-brown sometimes with the remains of the pale veil around the margin. The surface is smooth. The crowded gills are a golden-brown colour but become stained with dark brown spots with age. Spores orange. The stem is yellowish, paler above than below. It is cloaked in fibrous hairs and generally has a white base. Not edible. The yellowish flesh has a bitter taste.
Range and habitat: An extremely common and widespread species in Britain and northern Europe in suitable habitats. Grows on fallen twigs and logs of coniferous trees and sometimes birch.
Similar species: *C. junonius* is a larger species with a more fibrous surface. Some species of *Cortinarius* and *Pholiota* are superficially similar.

SLIPPER TOADSTOOL

Crepidotus mollis
Cap diameter: 2-6cm
Characteristics: The cap is a rounded, kidney-shaped bracket. The cap colour is pale brown when fresh but dries to almost pure white. Sometimes the cap may appear to have colour zones grading from pale-brown to white and is tiered. The surface has a gelatinous coating. The gills are crowded and are pale-brown at first, becoming darker with age. The spores are clay brown in colour. The stem is almost absent, often indistinguishable from the cap. Edibility unknown and so best avoided.
Range and habitat: A widespread and locally common species in Britain and northern Europe. Grows as horizontal brackets on the rotting logs and stumps of deciduous trees.
Similar species: *C. applanatus* has a more rounded, shell-shaped outline to the cap and lacks the gelatinous coating of *C. mollis*. *C. luteolus* has a creamy-yellow cap. Both species are uncommon and grow on decaying deciduous wood.

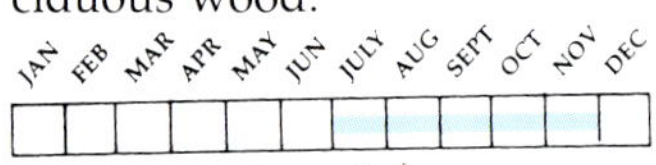

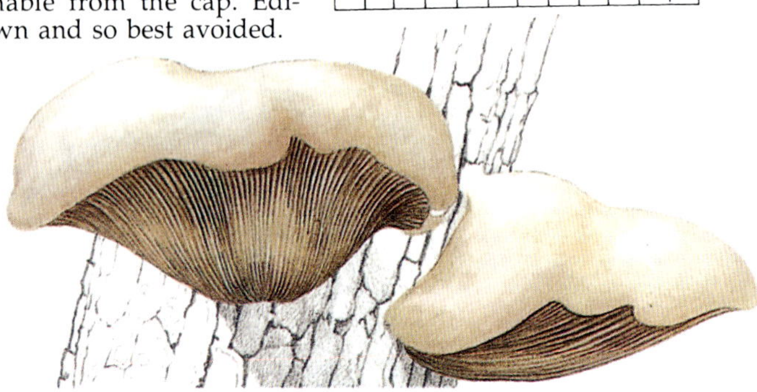

GYMNOPILUS JUNONIUS

(=*G. spectabilis*)

Cap diameter: 5-15cm

Height: up to 12cm

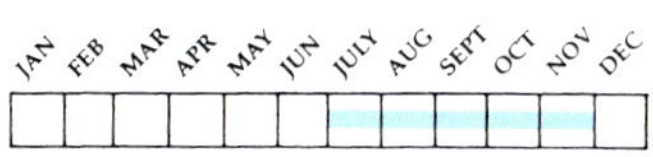

Characteristics: The cap is hemispherical at first but expands and flattens with age, occasionally with a slight umbo. The cap colour is a rich golden-brown and the surface is covered with fibrous scales forming radial streaks. The gills are orange at first, becoming rusty-brown with age. Spores orange. The stem is yellowish, sometimes stained and streaked darker, and is occasionally swollen in the lower half. The pronounced ring becomes stained orange-brown with the spores and often disintegrates. Not edible. The flesh is yellowish and is bitter.

Range and habitat: A common and widespread species. Grows in large, dense clusters at the base of deciduous trees or on logs or stumps.

Similar species: A striking and almost unmistakable species.

GALERINA MUTABILIS

(=*Keuhneromyces mutabilis*)

Cap diameter: 4-6cm

Height: up to 8cm

Characteristics: The cap is conical at first but becomes flattened with age, showing a shallow umbo. When fresh, the cap colour is a rich brown but as the fungus dries the colour fades to yellow-brown. The umbo and the margin usually retain the richer coloration producing a bi- or tricoloured pattern. The gills are pale-brown becoming darker with age. Spores ochre. The stem bears a prominent, brown ring. Above this the stem is pale and smooth; below, it is darker brown and scaly. Edible and considered good by many. Avoid due to confusion with similar inedible or poisonous species.

Range and habitat: Common and widespread. Grows in large clumps on the rotting stumps of deciduous trees.

Similar species: *G. marginata* (=*unicolor*) grows on conifer stumps. The bicoloured cap is dull brown in the centre with a buff margin. The stem is pale above the ring and darker below. Common and widespread.

ENTOLOMA CLYPEATUM

Cap diameter: 3-8cm

Height: up to 6cm

Characteristics: The cap is domed with a slight umbo when young. As it ages, the cap expands and becomes flattened with a low but pronounced umbo. The cap surface is often radially streaked and the margin becomes wavy and splits. The gills are grey at first but soon become flesh-pink. Spores pink. The stem is whitish but often streaked darker. Edibility uncertain and so best avoided since this species can be confused with other poisonous species of Entoloma. The flesh is dirty white with a mealy smell.

Range and habitat: A common and widespread species. Grows in grassy hedgerows, generally under Blackthorn or Hawthorn bushes.

Similar species: *E. sinuatum* has a grey-buff cap, flesh-pink gills when mature and white stem and flesh. It is a woodland and hedgerow species and is poisonous. *E. rhodopolium* has an olive-brown cap, flesh-pink gills when mature and whitish stem and gills. Common in grassy woodlands.

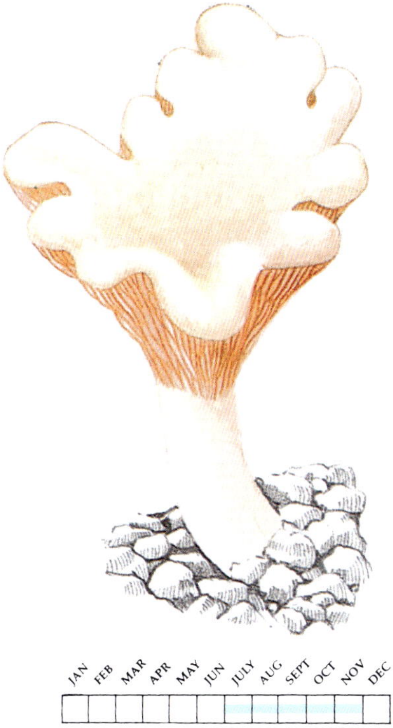

THE MILLER
Clitopilus prunulus
Cap diameter: 3-10cm
Height: up to 15cm
Characteristics: At first the cap is conical but with age it expands, eventually becoming funnel-shaped. The margin is wavy and the overall cap shape is often irregular and distorted. The colour is creamy-white and the texture of both the cap and the stem is that of kid leather. The gills are decurrent and crowded, white at first but becoming flesh-pink with age. Spores pink. The stem is the same colour as the cap and is often off-centre. Edible and good. The flesh is white and has a strong mealy smell.
Range and habitat: A common and widespread species in the region, found along grassy woodland rides and woodland margins.
Similar species: *C. prunulus* could be mistaken for small, pale species of *Clitocybe*, some of which are inedible or even poisonous. Most *Clitocybe* have a more regular funnel-shaped cap than the often distorted appearance of *C. prunulus*.

TRICHOLOMA FULVUM
Cap diameter: 4-8cm
Height: up to 8cm
Characteristics: The cap is broadly flattened with a shallow umbo. The general cap colour is dull orange-brown to tan with the centre rather darker than the margin. The cap surface bears faint radial streaks. The gills are yellowish-buff and acquire dark-brown spots with age. Spores white. The stem is the same colour as the cap and has a fibrous texture. Edible but not worth bothering with. The flesh is off-white in the cap and yellowish in the stem.
Range and habitat: A common and widespread species throughout the region. Grows in deciduous or mixed woodland, usually in association with Birch trees.
Similar species: *T. ustale* has a 4-8cm diameter, deep reddish-brown cap, darkening towards the centre. The stem is pale brown, darkening and becoming slightly scaly towards the base. Local and generally rather scarce. Grows in deciduous woodlands, especially under Beech.

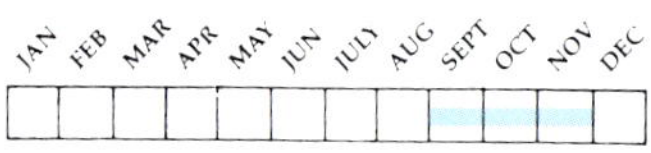

SULPHUR TOADSTOOL

Tricholoma sulphureum

Cap diameter: 4-8cm

Height: up to 6cm

Characteristics: A rather distinctive species. The cap is at first rounded and button-shaped but later expands and becomes flat with a shallow umbo. The cap colour is bright sulphur-yellow sometimes darkening towards the centre. The gills and stem are also bright sulphur-yellow, the stem being rather fibrous. Spores white. Not edible. The sulphur-yellow flesh tastes of meal and smells of coal gas.

Range and habitat: A widespread species throughout the range but seldom abundant. Grows under deciduous trees and but occasionally found under conifers.

Similar species: *T. sejunctum* has an often irregularly shaped olive-yellow cap, darkening towards the centre. The gills and stem are whitish. It is rather scarce and grows in deciduous woodland. *T. flavovirens* is a Scottish species found under conifers. It is yellow throughout and has a slightly scaly cap centre.

SOAP-SCENTED TOADSTOOL

Tricholoma saponaceum

Cap diameter: 3-10cm

Height: up to 12cm

Characteristics: A variable species whose cap is rounded at first but expands and flattens with age. The shape often becomes distorted and deep cracks and splits develop. The cap colour ranges from grey-buff to rich brown to almost black, generally rather paler around the margin. The gills are wavy, widely spaced and pale yellowish-buff, sometimes peppered with dark spots. Spores white. The stout stem is whitish often tinted with pink. Edible but not worth considering. The flesh is white, tinged with pink. It smells of soap and has a bitter taste.

Range and habitat: Widespread and sometimes locally abundant. It grows in deciduous and coniferous woodland.

Similar species: *T. argyraceum* has a 4-8cm diameter greyish-brown cap, usually with a slightly scaly appearance. The stem and gills are whitish and the flesh smells mealy. Uncommon under Beech and conifers.

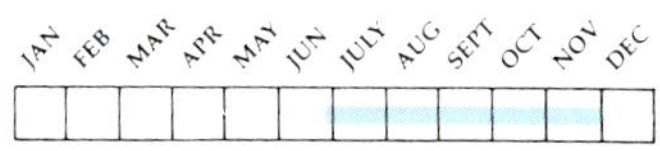

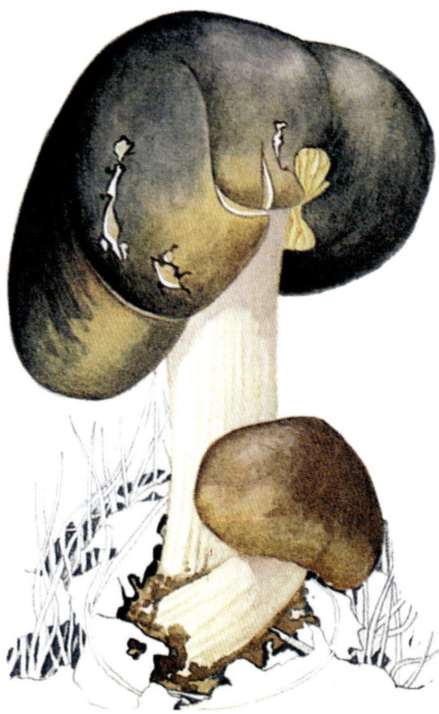

TRICHOLOMA USTALOIDES

Cap diameter: 5-10cm
Height: up to 10cm
Characteristics: The cap is hemispherical at first but expands to a broad dome with age. The cap sometimes cracks or splits with age and the margin becomes crinkly or wavy and slightly inrolled. The cap colour is a rich tan or chestnut- brown, darkening towards the centre and the surface is sticky and glistening. The gills are an off-white colour but gradually become marked with dark brown spots. Spores white. The stem is slightly tapering and is whitish at the apex with brown, scaly fibres becoming more apparent towards the base. Not edible. The flesh is white and has a mealy smell and a bitter taste.
Range and habitat: A widespread species but seldom numerous. Grows in deciduous woodland throughout Britain and northern Europe.
Similar species: *T. fracticum* is a similar species but usually grows in coniferous woodland. See also *T. ustale,* mentioned under *T. fulvum.*

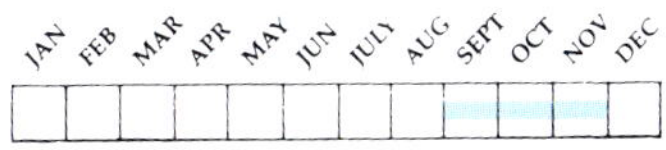

LYOPHYLLUM DECASTES

Cap diameter: 5-10cm
Height: up to 10cm
Characteristics: At first the cap is a rounded conical shape but it expands and flattens with age. The margin is wavy and the general shape is often irregular. The cap colour is grey-brown to muddy-brown. The surface is radially streaked and sometimes dimpled. The gills are dirty white. Spores white. The stem is white, darkening towards the base. It is tough and fibrous is often curved due to its growing position. Edible and considered worth eating by some people. The flesh is white.
Range and habitat: A common and widespread species throughout Britain and northern Europe. Grows in dense clusters, stems fused towards the base, along woodland rides and paths.
Similar species: *L. loricatum* is similar but has an orange-tan cap. It grows in deciduous woodland and is much less common than *L. decastes. L. connatum* is a pure white species. It is rather scarce and grows along grassy, woodland rides.

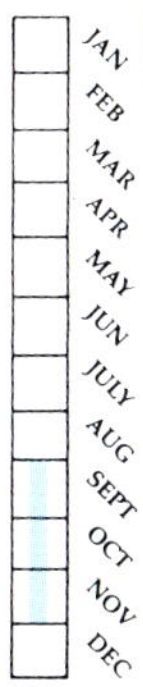

Calocybe gambosa

SAINT GEORGE'S MUSHROOM

Calocybe gambosa

Cap diameter: 5-15cm

Height: up to 15cm

Characteristics: The name derives from its appearance around St George's Day, 23rd April. The cap is domed at first but expands and flattens, the margin sometimes becoming cracked or split. The cap colour is creamy-white, often rather paler around the margin. The crowded gills are off-white and the stout stem is white, blotched with creamy-yellow. Spores white. Edible and good. The flesh is white and has a mealy smell and taste.

Range and habitat: A widespread species, sometimes locally common in suitable habitats. Grows in fields, meadows, downland and woodland edge.

Similar species: The appearance and time of year are good indicators to this species' identity. However, beware young stages of *I. patouillardii* whose flesh and surface bruise red. *Tricholoma columbetta* is an autumn woodland species. All parts are creamy white and the stem and cap are stained and streaked orange-brown.

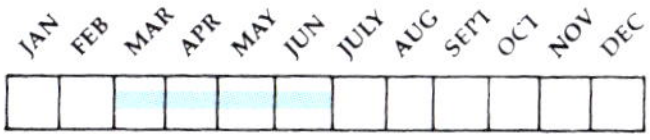

MELANOLEUCA MELALEUCA

Cap diameter: 4-8cm

Height: up to 8cm

Characteristics: At first, the cap is rather domed but it soon expands and acquires a distinctive shape: broadly flattened, depressed towards the centre with a rounded, shallow umbo. The cap colour is dark brown when fresh but drying pale greyish-buff. As it dries, different colour zones sometimes highlight the umbo and depressed cap centre. The gills are yellowish-cream. Spores cream. The stem has a bulbous base and is off-white streaked and stained with darker fibres. Edible but best avoided.

Range and habitat: A common and widespread species throughout the region. Grows along forest rides and paths and in fields bordering woodland.

Similar species: *M. grammopodia* is similar in appearance but is a scarcer species. It has a grey-brown or coffee-coloured cap and a whitish stem. Grows along grassy woodland rides and in fields. Sometimes found in large rings.

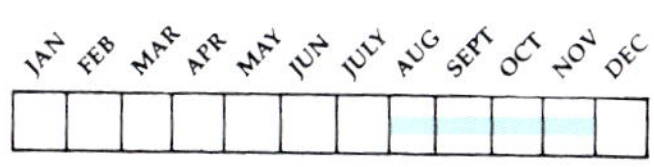

LEPISTA IRINA

Cap diameter: 5-10cm
Height: up to 10cm
Characteristics: The cap is domed and hemispherical at first but expands and flattens with age. The margin is slightly wavy in older specimens and often inrolled. The cap colour is clay to buffish-brown becoming more tanned towards the centre. The crowded gills are pale brown, acquiring a reddish tint with age. Spores pink. The stem is slightly tapered and has a somewhat bulbous base. It is an off-white colour and is streaked and lined with darker fibres. Edible and considered good by many people. The flesh is white and tastes and smells of perfume or violets.
Range and habitat: A widespread but rather local species in Britain and northern Europe. Grows in deciduous woodlands, often along paths.
Similar species: *L. irina* is superficially similar to other, more common, species of *Lepista* found in Britain. However, it lacks the violet or purple tints found on other species such as *L. saeva* and *L. nuda*.

WOOD BLEWITS

Lepista nuda
Cap diameter: 5-10cm
Height: up to 10cm
Characteristics: At first the cap is domed and conical but it expands and flattens sometimes becoming depressed in the centre. The cap colour is brown, sometimes tinged with red or violet when young. The surface is smooth and has a waxy feel to it. When young, the gills are an extremely attractive, intense bluish-purple colour but fade to buff with age. Spores pink. The stem is bluish-white, sometimes streaked and lined darker. The stem has a slightly bulbous base. Edible and delicious although causes stomach upsets in some people. The flesh is a delicate lilac colour.
Range and habitat: A common and widespread species throughout the region. Grows in deciduous woodlands, hedgerows and gardens among fallen leaves.
Similar species: *L. sordida* is a smaller, scarcer species also growing in woodland. The cap, gills and stem are generally paler.

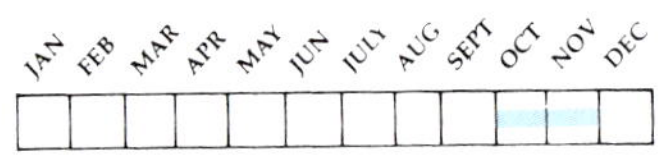

FIELD BLEWITS

Lepista saeva

Cap diameter: 5-10cm
Height: up to 8cm
Characteristics: The cap is domed and rounded at first but expands and flattens with age. In old specimens, the cap may become depressed in the centre and develop a wavy margin. When it grows in small clumps, the cap shape may be irregular and the stem off-centre. The cap colour is pale-brown or buff becoming darker towards the centre and the surface is dry and smooth. The gills are buff coloured. Spores pink. The stem is extremely bulbous at the base. It is whitish often tinged and veined heavily with intense lilac. Edible and delicious but it may cause stomach upsets in some people. The flesh is off-white.
Range and habitat: An extremely common and widespread species throughout the region. Grows in pastures, meadows and sometimes in mature gardens.
Similar species: Similar to *L. nuda* and *L. irina* but note lilac coloration on the stem and not on the gills or cap.

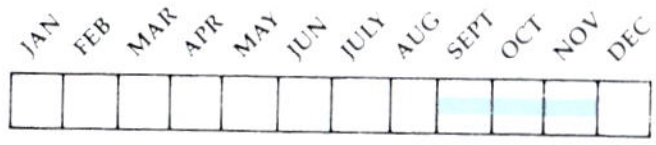

NYCTALIS PARASITICA

(=*Asterophora Parasitica*)

Cap diameter: 1-2cm
Height: up to 3cm
Characteristics: The cap is conical or bell-shaped at first but expands with age. The cap is delicate and white with slightly striated margins. The surface is smooth and silky. The gills are whitish and become brown and powdery with age. Spores buff. The stem is white, often twisted or curved, becoming darker with age. Not edible. The flesh is brown and has a revolting, rich mealy smell.
Range and habitat: An intriguing species which grows only on old and decaying specimens of certain *Lactarius* and *Russula* species. Not surprisingly, this species has a rather local distribution being found only in suitable woodlands where its hosts also grow.
Similar species: This species' appearance and habitat are unique. However, *A. lycoperdoides* grows on rotting specimens of *Russula nigricans*. The brown cap surface has a powdery, mealy texture, comprising special spores.

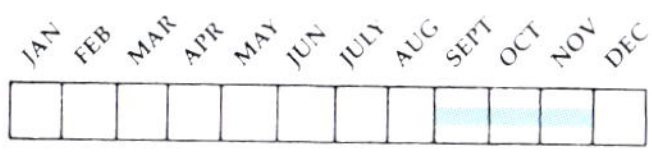

CLOUDED AGARIC

Clitocybe nebularis

Cap diameter: 5-15cm
Height: up to 14cm
Characteristics: At first the cap is domed but it expands and flattens with age. Eventually the cap becomes depressed in the centre with an inrolled margin. The cap colour is pinkish-brown to grey-brown, darkening towards the centre. The surface is smooth, sometimes with a powdery bloom. The gills are decurrent and creamy-yellow. Spores cream. The off-white stem is sometimes flushed with the cap colour. It is tapering and has a slightly swollen base. May be edible but best avoided because it has caused upsets in some people. Flesh white.
Range and habitat: A common and widespread species throughout the region. Grows in deciduous woodlands often in large groups or even rings.
Similar species: *C. inornata* is smaller and grows amongst leaf litter in both deciduous and mixed woodlands. It is pale-brown or buff in colour with the centre of the cap darkening. Less common than *C. nebularia.*

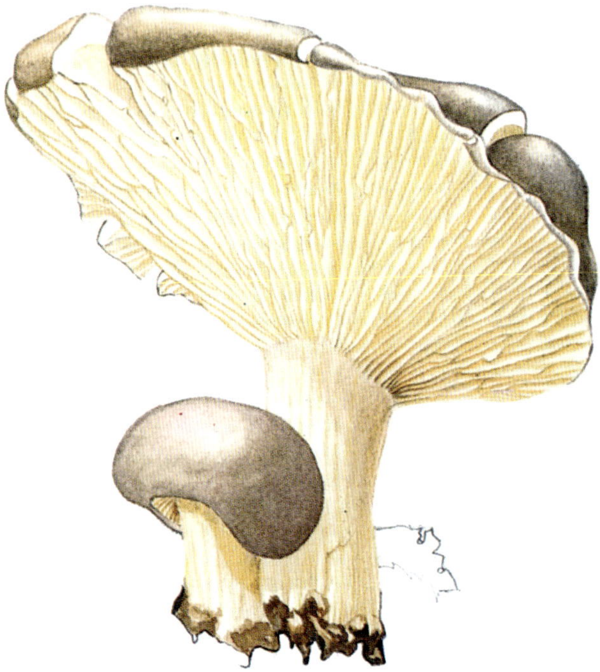

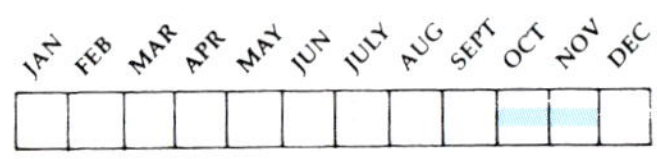

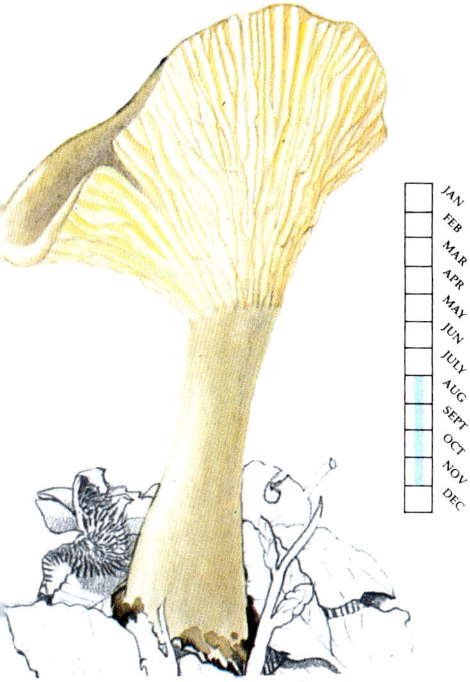

CLUB-FOOT

Clitocybe clavipes

Cap diameter: 4-8cm

Height: up to 8cm

Characteristics: At first the cap is domed but it expands and flattens, developing a slight umbo with age. The surface is smooth and the cap colour is grey-brown, sometimes tinted buff. The gills are decurrent and creamy-yellow in colour. Spores white. The stem is off-white with silky hairs and has a pronounced bulbous base from which the English name derives. Not edible. The flesh is whitish, becoming stained yellow in the stem. It has a fruity smell.

Range and habitat: A common and widespread species throughout the region. Grows in both deciduous and coniferous woodland amongst leaf litter.

Similar species: *C. phyllophila* has a slightly funnel-shaped cap with a wavy margin which is creamy-white to flesh-pink. The stem and gills are buff tinged with flesh-pink. A common and widespread species growing in deciduous woodland.

TAWNY FUNNEL CAP

Clitocybe flaccida (Lepista inversa)

Cap diameter: 5-10cm

Height: up to 10cm

Characteristics: The cap is slightly domed at first but soon expands and becomes funnel shaped with age. The centre is depressed without an umbo and the margin is slightly inrolled with occasional splits. The colour is from tawny-buff to orange-brown, darkening towards the centre, and the surface has the texture of soft leather. The gills are decurrent and buff coloured. Spores pink. The stem is buff-brown with a slightly swollen base. Edible but not recommended. The flesh is buff-brown in colour.

Range and habitat: A common and widespread species throughout the region which is sometimes locally abundant. Grows amongst leaf litter in both deciduous and coniferous woodland.

Similar species: *C. sinopica* has tan-coloured cap and stem with whitish gills. Rather uncommon and grows in woodland soils, particularly areas which have been recently burnt.

COMMON FUNNEL CAP

Clitocybe infundibuliformis

Cap diameter: 4-8cm

Height: up to 9cm

Characteristics: The cap is slightly domed at first but soon expands and becomes funnel-shaped with age. The centre is depressed and without an umbo and the margin is slightly wavy and occasionally splits. The colour is yellow-buff, sometimes tinged flesh-pink. The surface is smooth. The gills are decurrent and creamy-white in colour. Gills white. The tough, slender stem is slightly swollen at the base and is creamy-buff in colour. Edible and considered good by some people. The flesh is buff-white and has a sweet, slightly fruity smell.

Range and habitat: A common and widespread species throughout Britain and northern Europe. Grows in deciduous woodland among fallen leaves, and along grassy woodland rides and paths. Also found on heathland.

Similar species: *C. costata* is similar in size and shape but buff-tan in colour. Grows under conifers.

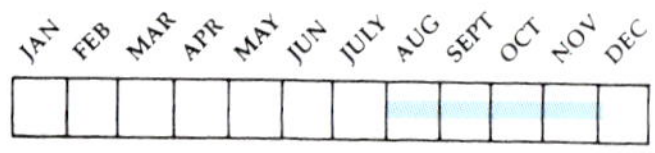

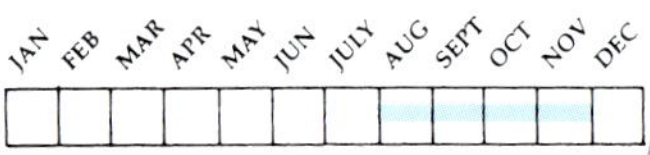

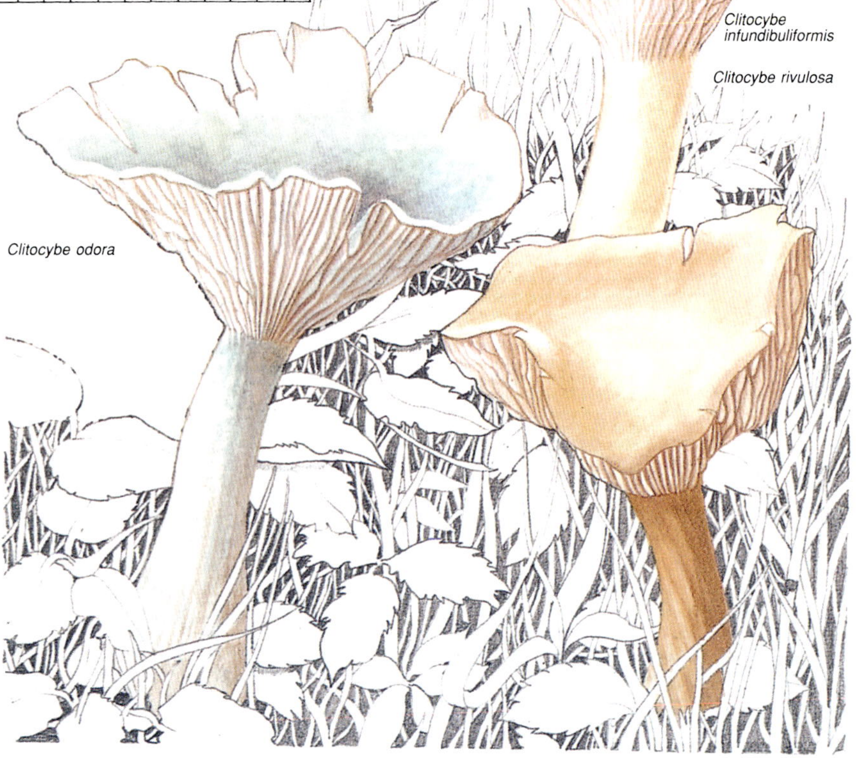

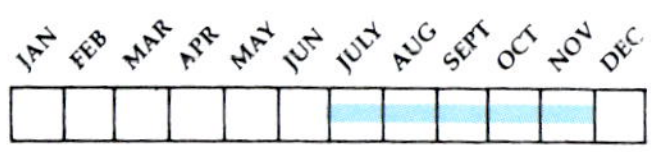

CLITOCYBE RIVULOSA
Cap diameter: 3-6cm
Height: up to 6cm
Characteristics: The cap is slightly domed at first but soon expands and flattens. May become slightly funnel-shaped with age. The cap colour is creamy-buff and the surface is smooth. In older specimens, the margin is wavy and slightly inrolled. The gills are decurrent and creamy-white in colour. Spores white. The tough, twisted stem is buff-brown in colour and woolly at the base. Deadly poisonous. The flesh is whitish.
Range and habitat: A common and widespread species throughout the region. Grows in grassy places such as roadside verges, woodland rides, garden lawns and meadows.
Similar species: This deadly species has a superficial resemblance to the edible *C. infundibuliformis*. Avoid eating if in any doubt about the identity. *C. dealbata* is another poisonous species found in grassy habitats. It is a slightly funnel-shaped cap with decurrent gills and all parts are a dirty whitish-buff.

ANISEED TOADSTOOL
Clitocybe odora
Cap diameter: 4-8cm
Height: up to 7cm
Characteristics: The cap is domed at first but expands and flattens with age, developing a slight umbo. May become slightly funnel-shaped in older specimens, with a wavy, split margin. The colour is greenish-grey, tinged with buff in some specimens. The gills are slightly decurrent and pale greenish-white in colour. Spores white. The stem is a dirty greenish-white colour, slightly swollen and woolly at the base. Edible and good. The flesh is whitish and has a strong smell and taste of aniseed. Use sparingly because of strong taste.
Range and habitat: A common and widespread species throughout Britain and northern Europe. Grows among fallen leaves in deciduous woodland and along paths and rides.
Similar species: The strong aniseed smell and taste and distinctive appearance make this species difficult to mistake. *C. fragrans* also smells of aniseed. Smaller and pale-brown.

CLITOCYBE GEOTROPA
Cap diameter: 10-20cm
Height: up to 20cm
Characteristics: At first the cap is a flattened button-shaped dome with a distinct umbo and an inrolled margin. It expands and flattens with age and eventually becomes slightly funnel-shaped, often still retaining the umbo. The cap colour is yellowish-buff. The gills are crowded and decurrent and slightly paler than the cap. Spores white. The tough, rather slender stem is brownish-buff in colour. It is swollen and slightly woolly at the base. Edible and good. The flesh is white with a distinct smell of hay.
Range and habitat: A common and widespread species throughout Britain and northern Europe. Grows in deciduous woodland, often along rides and paths. Sometimes locally abundant and occasionally found growing in large rings.
Similar species: The closely related *Leucopaxillus giganteus* is larger with a relatively short stem. It is found in more open, grassy places.

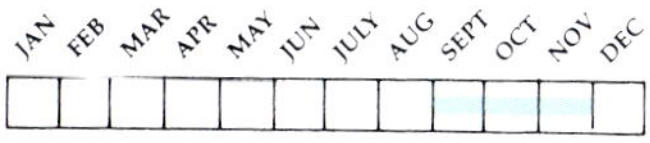

CLITOCYBE LANGEI
Cap diameter: 2-5cm
Height: up to 5cm
Characteristics: At first the cap is domed with a depressed centre. This expands and flattens with age still generally retaining the depressed centre. The cap colour is muddy-brown to orange-tan but as the fungus dries the centre becomes darker, contrasting with the pale margin. When wet, the cap margin is striate. The gills are decurrent and greyish-brown in colour. Spores white. The stem is tough and fibrous with a slightly woolly base. Not edible. The flesh is dirty white in colour.
Range and habitat: A widespread but rather uncommon species throughout the region, found in mixed and coniferous woodland. Often grows in association with bracken and birch.
Similar species: *C. vibecina* is very similar in size and shape. The colour is buff-brown, darkening towards the depressed centre of the cap. The margin is striate when wet but dries paler. *C. vibecina* grows in similar habitats to the previous species.

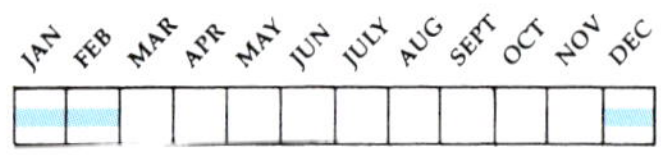

LEUCOPAXILLUS GIGANTEUS
Cap diameter: 10-30cm
Height: up to 15cm
Characteristics: The cap is a button-shaped flattened dome at first but this soon expands greatly. Eventually becomes funnel-shaped with an inrolled margin. The colour is creamy-white, sometimes stained darker buff towards the centre. The cap surface is rather downy and cracks and splits with age towards the centre. The gills are creamy-white and decurrent. Spores white. The stem is tough and stout and also creamy-white in colour. Edible and considered good by some people. The flesh is white with a mushroomy smell.
Range and habitat: A widespread species in Britain and northern Europe. Nowhere common but difficult to miss because of its size. Grows in grassy places such as meadows, roadside verges, woodland clearings and hedgerows. May form large rings.
Similar species: Difficult to mistake because of its size and shape. *Clitocybe geotropa* is smaller and has a relatively long, slender stem.

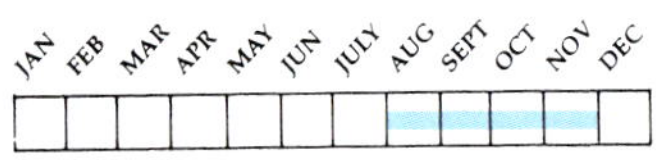

FALSE CHANTERELLE
Hygrophoropsis aurantiaca
Cap diameter: 3-6cm
Height: up to 7cm
Characteristics: At first the cap is a flat-topped funnel-shape but later it expands and becomes slightly depressed in the centre. The surface is downy and the margin is often inrolled and split. The cap colour is orange-yellow, sometimes paler. The gills are decurrent and paler orange-yellow. Spores white. The stem is the same colour as the cap. Doubtfully edible and best avoided since it may cause stomach upsets. The flesh is orange-yellow and smells mushroomy.
Range and habitat: A common and widespread species throughout Britain and northern Europe. Grows in both mixed and coniferous woodland often in association with birch on heathland.
Similar species: The Chanterelle, *Cantharellus cibarius*, is a superficially similar species which is a paler yellow colour. The gills are absent, replaced by gill-like ridges and the flesh smells strongly of apricot.

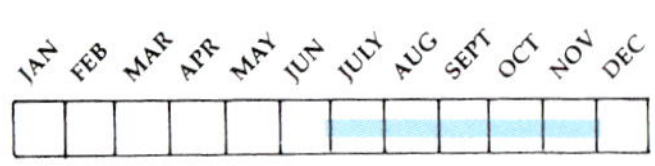

Leucopaxillus giganteus
Hygrophoropsis aurantiaca
Clitocybe langei

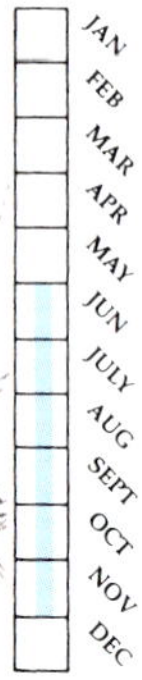

DECEIVER

Laccaria laccata

Cap diameter: 2-5cm
Height: up to 10cm
Characteristics: The English name reflects this species' very variable nature which may make initial identification difficult. The cap is domed at first but expands and flattens with age. The margin may be wavy and split and the centre depressed in old specimens. The cap colour is orange-tan, drying paler. The gills are widely spaced and pinkish-orange, becoming covered in powdery white spores with age. The tough stem is the same colour as the cap and often twisted. Edible but probably best avoided if identification is in any doubt. The flesh is reddish-orange.
Range and habitat: An extremely common and widespread species throughout the region. Grows in a wide variety of habitats from deciduous woodlands, to grassy hedgerows and heathlands.
Similar species: *L. proxima* is a larger, stouter species with a more rounded cap. It grows in a variety of wooded and grassy habitats.

AMETHYST DECEIVER

Laccaria amethystea

Cap diameter: 2-5cm
Height: up to 10cm
Characteristics: A beautiful and distinctive species. The cap is domed at first but expands and flattens with age, sometimes with a slightly depressed centre. The cap colour is a deep purple or lilac when wet, drying to pale pinkish-lilac. The gills are deep purple but become dusted with white, powdery spores with age. The stem is the same colour as the cap and is covered in white hairs towards the base. Edible but not worth eating. The flesh is whitish-lilac.
Range and habitat: An extremely common species throughout the region. Grows in a variety of wooded habitats, preferring shady deciduous areas.
Similar species: *L. bicolor* is rather similar to the Deceiver, *L. laccata,* with a buff-tan cap and reddish gills. The tough and twisted stem is buff-tan at the apex becoming covered in lilac or purple downy hairs towards the base. Rather uncommon. Grows in mixed woodland and under conifers.

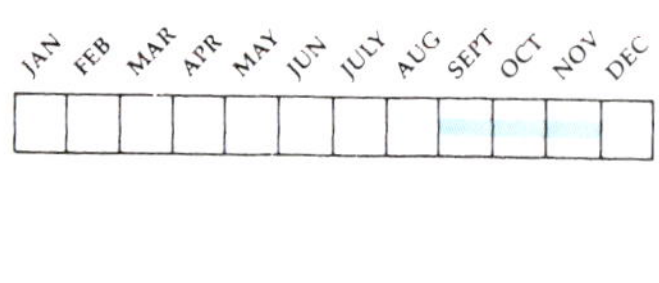

OMPHALINA (GERRONEMA) ERICETORUM

Cap diameter: 0.5-2cm
Height: up to 2.5cm
Characteristics: At first the cap is slightly domed with a depressed central area. It expands and becomes funnel-shaped with age, retaining the depressed centre. The cap colour is olive-brown to tan. The surface bears radial grooves and the margin is sometimes irregular and splits with age. The gills are decurrent and slightly paler than the cap. Spores white. The slender stem is the same colour as the cap, sometimes darkening towards the apex. The base of the stem is slightly woolly. Not edible. The flesh is yellowish.
Range and habitat: A common and widespread species in Britain and northern Europe. Grows on peaty soils in woodland and on heathland. Sometimes locally abundant in suitable areas and may form large groups.
Similar species: *O. pyxidata* is similar but is reddish-brown. It grows in grassy areas such as lawns and woodland paths and is a widespread species.

JAN	FEB	MAR	APR	MAY	JUN	JULY	AUG	SEPT	OCT	NOV	DEC

BUTTERY TOUGH SHANK; BUTTER CAP

Collybia butyracea
Cap diameter: 4-8cm
Height: up to 8cm
Characteristics: At first the cap is domed with a pronounced umbo. It expands and flattens with age, still retaining the umbo. The cap colour is olive-brown to reddish-brown. As it dries, the cap colour fades but the umbo and margin usually remain darker. The result is often three concentric zones of colour. The cap surface is greasy or buttery when moist, hence the English name. The gills are greyish white. Spores white. The tapering stem is buff-brown and has a swollen base, covered in white hairs. Edible but not recommended. The flesh is off-white and has a rancid smell.
Range and habitat: A common and widespread species throughout Britain and northern Europe. It grows in leaf-litter in a wide range of woodland habitats including under both coniferous and deciduous trees.
Similar species: The greasy feel to the cap makes this species easy to identify.

JAN	FEB	MAR	APR	MAY	JUN	JULY	AUG	SEPT	OCT	NOV	DEC

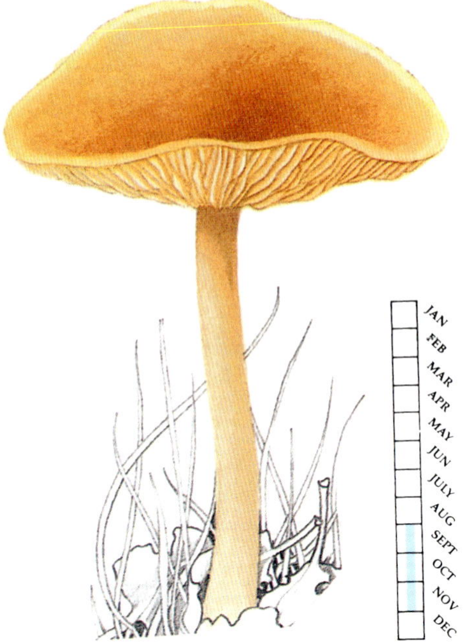

OAK TOUGH SHANK

Collybia dryophila

Cap diameter: 2-4cm
Height: up to 5cm
Characteristics: At first the cap is a flattened dome shape but this later expands and flattens further. The mature cap has a wavy margin and is almost always an uneven shape. The cap colour is extremely variable from buffish-white tinged with flesh-pink to orange-tan, darkening towards the centre. The gills are yellowish-white. Spores white. The tough, fibrous stem is flexible and orange-tan in colour, darkening towards the base. The base is slightly swollen and hairy, although not especially so. Edible but not worth considering because of small size and tough texture.
Range and habitat: A common and widespread species in Britain and northern Europe. Sometimes locally abundant. Grows in deciduous woodlands as well as on heathlands. It has a preference for oak woodland.
Similar species: The Wood Woollyfoot, *C. peronata*, is superficially similar but has an extremely hairy stem base.

SPINDLE SHANK

Collybia fusipes

Cap diameter: 3-6cm
Height: up to 10cm
Characteristics: The cap is domed at first with a broad umbo and later expands and flattens, retaining the umbo. The surface is slightly sticky in wet weather and the cap colour is reddish-brown when wet. As it dries, the margin becomes paler brown. The gills are buffish-white. Spores white. The tough stem is pale-brown to tan, often darkening towards the base. It is often swollen in the middle and spindle-shaped, hence the English name. The surface is grooved and twisted and the base is often rooted and fused to other stems. Not edible. The flesh is tough and white.
Range and habitat: A widespread and common species. Grows in dense clumps at the base of deciduous trees and their stumps. Especially on Oak.
Similar species: *C. distorta* has a reddish-tan cap, white gills and a slender, pale-brown stem which is twisted and grooved. Rather uncommon. Grows under conifers.

WOOD WOOLLY-FOOT

Collybia peronata
Cap diameter: 3-6cm
Height: up to 7cm
Characteristics: The cap is slightly conical at first but expands and flattens with age, often bearing a broad umbo. Mature specimens may have the centre of the cap depressed and a rather wrinkled surface. The cap colour is orange-brown to dark-tan. The gills are orange-brown to rich-brown in colour. Spores white. The slender, tough stem is pale-brown and covered in dense, woolly hairs towards the base, hence the English name. Not edible. The flesh is off-white and has a bitter and peppery taste.
Range and habitat: A common and widespread species in Britain and northern Europe. It grows in leaf-litter in deciduous woodland, sometimes in clumps. Sometimes found under conifers.
Similar species: *C. fuscopurpurea* is similar to *C. peronata* but dark-brown in colour, drying paler. It is less common and grows in deciduous woodland, especially under Beech.

SPOTTED TOUGH SHANK

Collybia maculata
Cap diameter: 5-10cm
Height: up to 12cm
Characteristics: The cap is domed at first but soon expands and flattens. The cap surface is smooth and the margin often folds and warps as it matures producing a rather uneven shape. The cap colour is white at first but this becomes spotted and stained reddish-brown with age. Eventually, the cap may become almost completely brown. The gills are white but become spotted reddish-brown with age. Spores pinkish. The tough, slender stem is white becoming spotted and blotched reddish-brown. Often strongly rooted. Not edible. The flesh is white and bitter.
Range and habitat: A common and widespread species. Grows mainly in deciduous woodland. May form large rings. Often associated with bracken.
Similar species: The white colour, staining and bruising red, are good clues to this species' identity. *C. confluens* has a brown, hairy stem and a buff cap which does not stain.

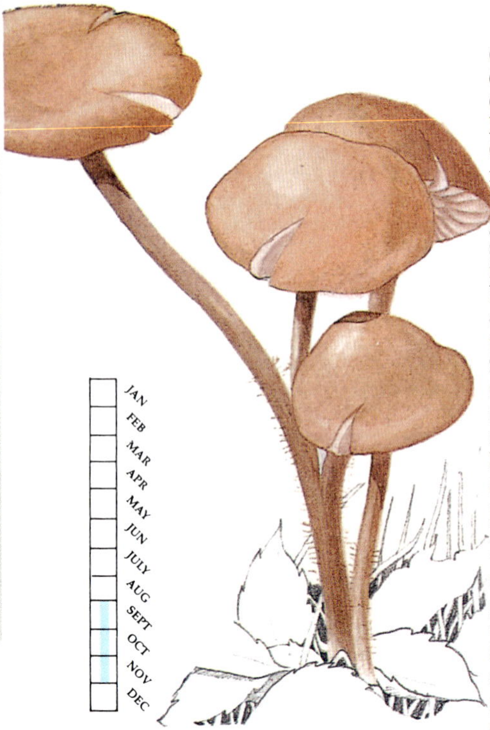

CLUSTERED TOUGH-SHANK

Collybia confluens
Cap diameter: 3-6cm
Height: up to 10cm
Characteristics: At first the cap is a rounded, conical shape but this expands and flattens with age. The margin may warp or split in older specimens. The colour is reddish-buff in fresh specimens but this fades to buffish-brown as the fungus dries. The centre generally remains darker than the margin. The gills are pinkish-buff. Spores white. The tough, slender stem is reddish-brown and is covered in fine, hairs. Within clumps, several stems are often fused together from the base for part of their length. Edible but not worth considering. The flesh is pinkish-white and tough.
Range and habitat: A common and widespread species. Grows in leaf-litter in deciduous woodland. Often grows in large clumps and may form extensive rings on the woodland floor.
Similar species: *C. erythropus* has a slender, red stem and buff cap and gills. It is uncommon and grows in clumps in deciduous woodland.

HORSE-HAIR FUNGUS

Marasmius androsaceus
Cap diameter: 0.5-1cm
Height: up to 8cm
Characteristics: At first the cap is domed with a pronounce depression in the centre. The membranous cap expands with age, still retaining the depression. The surface is wrinkled and is streaked with radial grooves. The cap colour is reddish-tan to pinkish-brown. The reddish-brown gills are widely spaced and connected to a central collar rather than to the stem. Spores white. The tough and slender stem is black and hair-like. Not edible.
Range and habitat: A common and widespread species in Britain and northern Europe. Grows on the fallen twigs and needles of conifers. Also found on the dead twigs of heather. Locally abundant in suitable habitats.
Similar species: The unrelated *Baeospora myosura* is much stouter fungus with a tan-coloured cap and pale-brown gills and stem. It grows on decaying conifer cones and is usually solitary.

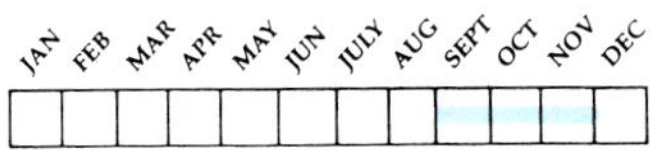

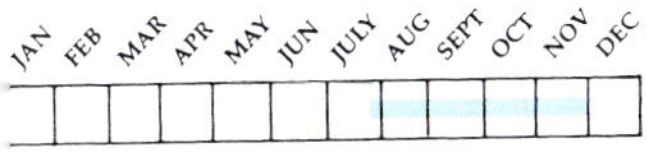

FAIRY-RING TOADSTOOL

Marasmius oreades

Cap diameter: 3-6cm
Height: up to 10cm
Characteristics: At first the cap is domed with a pronounced umbo. The cap expands and flattens with age but retains the umbo. The margin is often grooved. The cap is tan coloured when wet. In dry specimens the cap fades to buff or pale-brown except for the umbo which remains darker tan. The gills are buff-brown. Spores white. The tough stem is pale brown and woolly towards the base. Edible but best avoided because of potential confusion with certain poisonous *Clitocybe* species. The flesh is whitish.
Range and habitat: A common and widespread species throughout the region. Grows in grassy places such as lawns, parks and meadows. Often grows in large rings, hence the English name.
Similar species: *M. alliaceus* has a long, slender stem which is brown and a pale-brown or tan cap and gills. Uncommon in beech woods. Smells strongly of garlic.

MARASMIUS ROTULA

Cap diameter: 0.5-1.5cm
Height: up to 10cm
Characteristics: The cap is an ornate and beautiful dome-shape with a markedly depressed centre. The surface is smooth and white and it is ridged with radial ribs. The margin is scalloped. The gills are widely spaced and attached not to the stem but to a cap collar situated at the stem apex. Spores white. The cap resembles a miniature parachute or umbrella when viewed from the underside. The gill colour is creamy-white. The stem is very slender and sinuous. It is black and shiny, becoming paler beneath the cap. Not edible.
Range and habitat: A common and widespread species in Britain and northern Europe. Grows on decaying twigs and roots on the woodland floor. Sometimes locally abundant on suitable twigs.
Similar species: *M. epiphyllus* is similar in appearance but smaller with an incredibly slender stem. Grows from rotting twigs and leaf petioles on the woodland floor. Rather uncommon.

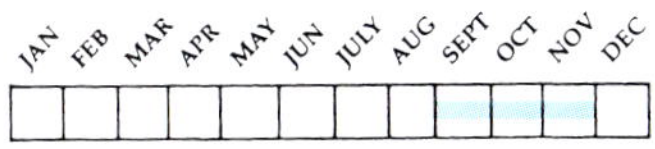

Flammulina velutipes

Tricholomopsis rutilans

Tricolomopsis platyphylla

VELVET SHANK
Flammulina velutipes
Cap diameter: 2-8cm
Height: up to 10cm
Characteristics: At first the cap is conical but it expands and flattens with age. The surface is rather shiny and sticky and is often undulating in older specimens. The cap colour is bright yellow or orange, often darkening towards the centre when wet. The gills are yellow or pale orange. Spores white. The tough stem is orange-brown, darkening towards the base. The stems are usually curved upwards from their growing point and are velvety towards the base, hence the English name. Edible and considered delicious by many people. The flesh is yellowish.
Range and habitat: A common and widespread species throughout Britain and northern Europe. Grows in dense clumps on dead and decaying stumps of deciduous trees, especially Elm.
Similar species: The bright orange cap, velvety stem and appearance in winter and early spring make this species readily identifiable.

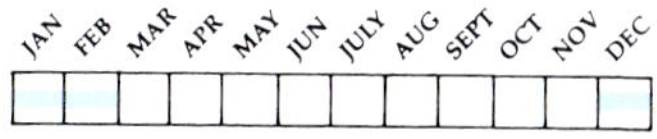

PLUMS AND CUSTARD
Tricholomopsis (Tricholoma) rutilans
Cap diameter: 4-10cm
Height: up to 10cm
Characteristics: A most distinctive and attractive species. The cap is domed at first but later expands and flattens with age. The margin is rather undulating in older specimens and the cap sometimes has a slight umbo. The cap is a yellowish colour but is covered in a dense layer of overlapping reddish-brown scales and tufts, more densely packed and darker towards the centre. The gills are orange-yellow. Spores white. The stem is a brownish-yellow colour and is covered in reddish-brown scales and striations. Doubtfully edible and so best avoided.
Range and habitat: A widespread and often abundant species in the right habitat. Grows on, and in association with, conifer stumps.
Similar species: *T. decora* also grows on conifer stumps but is much less common. All parts are yellowish-orange and the cap is covered in darker fibres and scales.

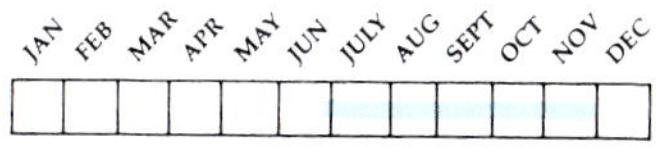

BROAD-GILLED AGARIC
Tricholomopsis (Collybia) platyphylla
Cap diameter: 6-12cm
Height: up to 12cm
Characteristics: The cap is domed at first but expands and flattens with age, the centre often becoming depressed. The cap is radially streaked with fibres and the surface sometimes splits and cracks in older specimens. The colour is greyish-brown, darkening towards the centre. The gills are widely spaced and reddish-buff in colour. Spores white. The stem is greyish-brown attached at the base by tough, string-like mycelial strands to the growing medium. Not edible. The flesh is white and has a bitter taste.
Range and habitat: A common and widespread species throughout the region. Grows in deciduous woodland, especially under Oak, attached to stumps, logs and buried wood.
Similar species: *Tricholoma virgatum* is rather uncommon and is found in both coniferous and deciduous woodland. The grey cap is radially streaked. Stem and gills whitish.

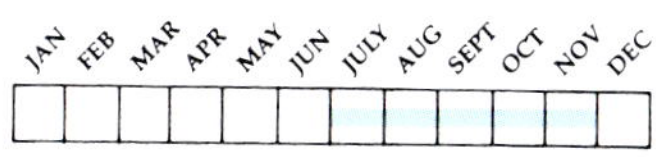

HONEY FUNGUS

Armillaria mellea

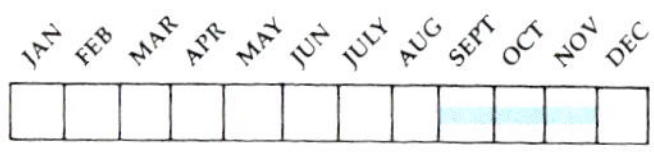

Cap diameter: 5-15cm
Height: up to 15cm
Characteristics: The cap is rounded at first but expands and flattens, sometimes with a shallow umbo. In mature specimens, the margin may be undulating. The cap colour is pale-brown or tan and the surface is covered dark brown scales, crowded towards the centre. The gills are yellowish. Spores cream. The stem, which bears a ring, is slightly bulbous towards the base. It is yellowish, darkening towards the base. Black, bootlace-like mycelial strands spread from the stem base under tree bark and through rotting timber. Not edible. Flesh whitish and bitter.
Range and habitat: A common and widespread species. Grows in large clumps on deciduous and coniferous trees. Attacks stumps and living trees eventually causing death of the host.
Similar species: *A. tabescens* grows on deciduous trees. It is orange-brown and the stem lacks a ring.

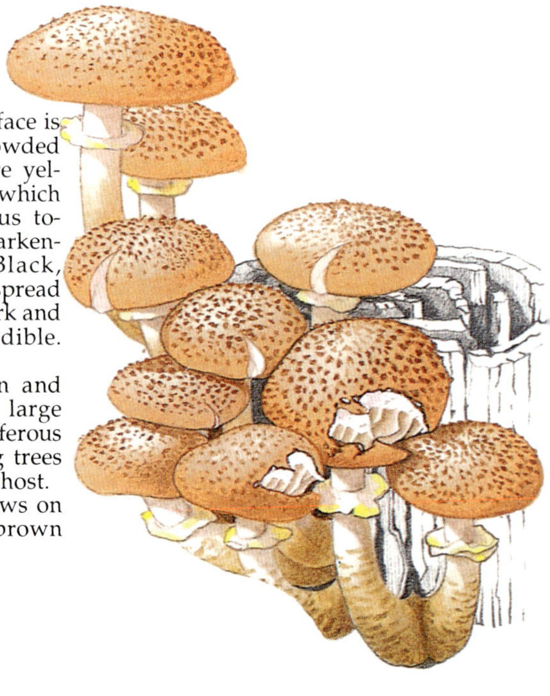

PORCELAIN FUNGUS

Oudemansiella mucida
Cap diameter: 4-8cm
Height: up to 10cm
Characteristics: At first the cap is domed but it later expands and flattens with age, often with a slight umbo. The cap is whitish and semi-translucent with a slimy, glistening surface. The gills are widely spaced and white. Spores white. The tough stem bears a delicate, sticky ring. The stem is whitish in colour, becoming slightly darker and scaly towards the base. Edible but not to be recommended since it is glutinous and slimy unless washed thoroughly.
Range and habitat: A common and widespread species in Britain and northern Europe. It grows almost exclusively on the trunks and branches of Beech trees. It generally occurs on damaged trees or on fallen branches and trunks. This species sometimes forms large clusters.
Similar species: Difficult to confuse with any other because of its white colour, translucent appearance and habitat.

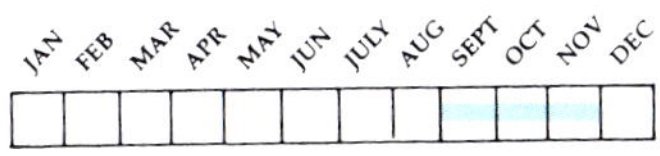

ROOTING SHANK

Oudemansiella radicata
Cap diameter: 4-8cm
Height: up to 15cm
Characteristics: The cap is a rounded cone-shape at first but expands and flattens with age. Mature specimens often have the cap surface depressed in the middle with a raised, central umbo. The cap surface is slimy in wet weather with radial wrinkles and striations. The colour is pale-brown to tan, darkening towards the centre with a paler umbo. The gills are widely spaced and white. Spores white. The stem is white, staining yellowish-brown towards the base and has a long rooting portion buried beneath the ground. Edible but not worth considering. The flesh is whitish.
Range and habitat: A common and widespread species throughout the region. It is found in association with roots or buried timber in deciduous woodland.
Similar species: *O. longipes* is similar but has a dry, downy cap. The stem is also downy and all parts are pale brown in colour.

Mycena galericulata

ORANGE BONNET

Mycena acicula

Cap diameter: 0.25-1cm
Height: up to 5cm
Characteristics: An attractive species. The cap is hemispherical or conical in shape. The cap colour is bright orange, becoming paler towards the margin, and the surface is radially streaked and lined. The gills are widely spaced and are pale yellow with a white border. Spores white. The stem is slender and smooth and pale yellow in colour. The base is slightly hairy. Not edible. Flesh yellowish.
Range and habitat: A widespread species in Britain and northern Europe. Locally common in deciduous woodland. Grows in small groups on fallen twigs and branches.
Similar species: *M. adonis* is slightly larger with reddish-pink cap and gills and a pale pink stem. It grows on in coniferous woodland and is rather scarce. *Rickenella* (=*Mycena*; =*Omphalina*) *fibula* is somewhat similar. The orange cap is flattened with a depressed centre and the gills are decurrent. Grows among damp moss.

JAN	FEB	MAR	APR	MAY	JUN	JULY	AUG	SEPT	OCT	NOV	DEC

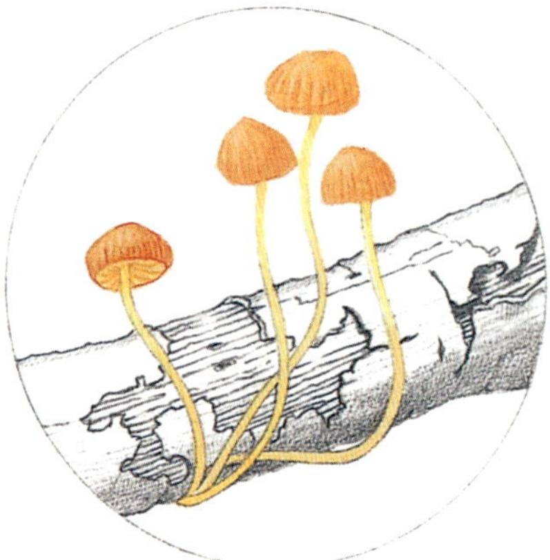

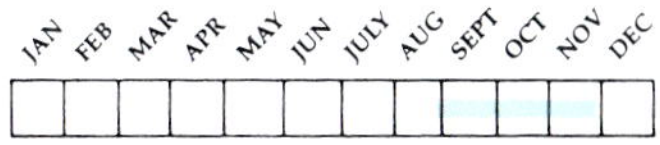

MYCENA ALCALINA

Cap diameter: 1-4cm
Height: up to 7cm
Characteristics: The cap is conical at first but expands and becomes bell-shaped with age with a pronounced umbo. The cap colour is grey-brown and the umbo is sometimes markedly paler. Some specimens of this species are much darker. The cap surface is smooth and, in wet weather, radially striated. The gills are white and widely spaced. Spores white. The stem is grey-brown, slender and smooth. Apparently edible but not worth considering. The flesh is white and smells strongly of bleach.
Range and habitat: A common and widespread species in Britain and northern Europe. It is found almost exclusively on stumps and buried timber of coniferous trees and grows in small, tufted clumps.
Similar species: *M. leptocephala* has a 1cm diameter conical, grey cap, grey gills and a slender, 5cm grey-brown stem. It grows among fallen leaves, pine needles and short grass and is a common and widespread species.

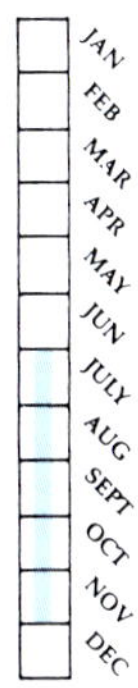

MYCENA EPIPTERYGIA

Cap diameter: 1-2cm
Height: up to 8cm
Characteristics: The cap is conical at first but sometimes becomes bell-shaped with age. The cap surface is very sticky and the colour is tan to yellow-buff, sometimes darker towards the centre. In wet weather, the cap margin is distinctly striated. The gills are whitish or pale-brown. Spores white. The long, slender stem is whitish becoming yellow towards the base and is sticky. Edible but not worth considering because of its small size and indifferent taste. The flesh is whitish.
Range and habitat: A common and widespread species throughout the region. Grows among pine needles in coniferous woodland and in association with bracken and ling on heathland and moorland.
Similar species: *M. viscosa* is a more robust species with a brown cap, darkening towards the centre and yellowish gills. The stem is yellow and it grows in coniferous woodland among pine needles and on stumps.

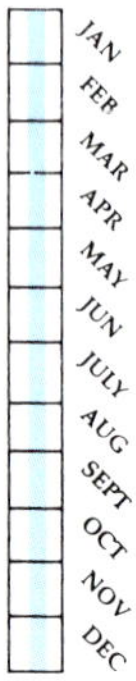

BONNET MYCENA

Mycena galericulata

Cap diameter: 2-5cm
Height: up to 10cm
Characteristics: The cap is conical at first but later expands with age and becomes bell-shaped with a distinct umbo. Some mature specimens have an undulating margin which splits and frays. The cap colour varies from grey-brown to tan with the umbo sometimes slightly darker. The cap margin is striated. The gills are white at first but gradually become flushed with pink. Spores white. The slender stem is long and greyish-brown in colour covered with white, downy fibres towards the base. Edible but not worth considering. Flesh white.
Range and habitat: Common and widespread in deciduous woodland. Grows in clumps on rotting stumps and fallen branches.
Similar species: *M. aetites* is smaller with a rather domed, radially striated cap. All parts are greyish-brown and it is a common species, growing in clumps in coniferous woodland.

MYCENA INCLINATA

Cap diameter: 1-3cm
Height: up to 10cm
Characteristics: The cap is conical at first but later expands and becomes bell-shaped with age with a pronounced umbo. The cap colour is grey-buff to reddish-brown, darkening towards the centre, and the surface bears radial lines and grooves. The margin often has a scalloped appearance and splits and frays in older specimens. The gills are whitish becoming flushed with pink. Spores white. The long, slender stem is reddish-brown or buff, darkening towards the base which is covered in white down. Not edible. Flesh is white and rancid.
Range and habitat: Common and widespread throughout the region. Grows in deciduous woodland, forming clumps on rotting Oak stumps and fallen branches.
Similar species: *M. maculata* has a flattened, bell-shaped cap which is grey-brown, becoming spotted and stained red with age. The gills and stem are also grey-brown staining red. Grows in clumps on stumps.

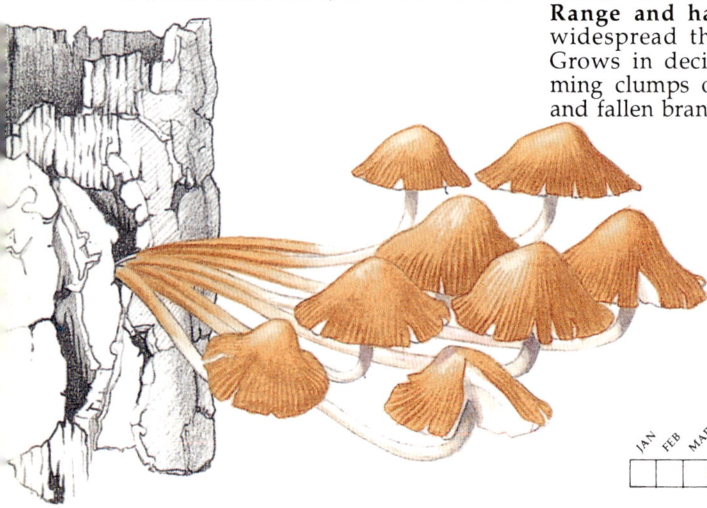

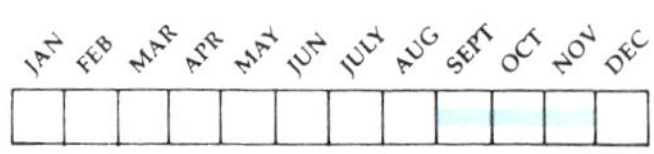

MYCENA POLYGRAMMA

Cap diameter: 2-4cm
Height: up to 10cm
Characteristics: The cap is conical at first but expands with age and becomes bell-shaped. The cap colour is grey-brown becoming buff or tan towards the central umbo. The broad margins of the cap are radially streaked with black lines. The margin may split or fray in older specimens. The widely-spaced gills are greyish-white sometimes becoming flushed with pink. Spores white. The stem is long and slender. It is greyish-brown and is often kinked at the rooting base. Not edible. The flesh is whitish.
Range and habitat: A common and widespread species throughout the region. It grows in deciduous woodland and is found on rotting logs and stumps and on buried timber.
Similar species: *M. vitilis* (=*filopes*) is a delicate species with a 2cm diameter cap. The flattened cap is grey-brown in colour and the stem is long and slender. It grows on buried timber in deciduous woodland and is common and widespread.

BLEEDING MYCENA

Mycena haematopus
Cap diameter: 1-3cm
Height: up to 10cm
Characteristics: The cap is conical at first but later expands and becomes bell-shaped sometimes with a prominent umbo. The cap colour is reddish-tan to muddy-brown, often paler towards the central umbo. In wet weather, the margin is striate and in mature specimens it may split and crack. The gills are white at first, tinged with red with age. Spores white. The stem is long, slender and reddish-brown in colour. When broken or damaged, it produces a blood-red latex, hence the English name. Edible but not worth considering. The flesh is red.
Range and habitat: A widespread species throughout the region which is locally common in deciduous woodland. It grows on rotting stumps.
Similar species: *M. crocata* is slightly smaller and greyish brown becoming stained and spotted orange-tan with age. When broken or damaged, the stem exudes orange latex. Rather uncommon in Beech woods.

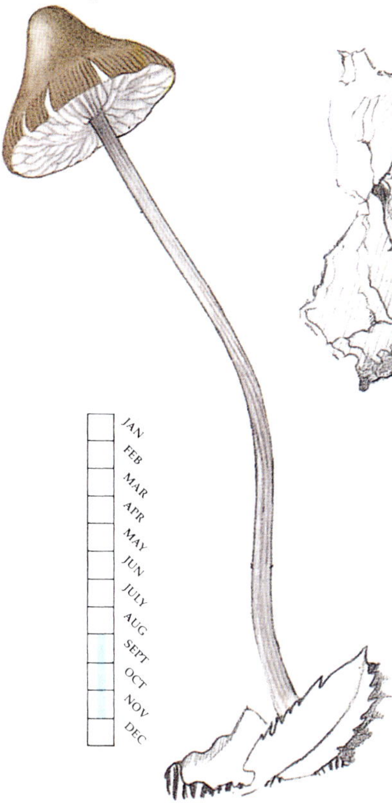

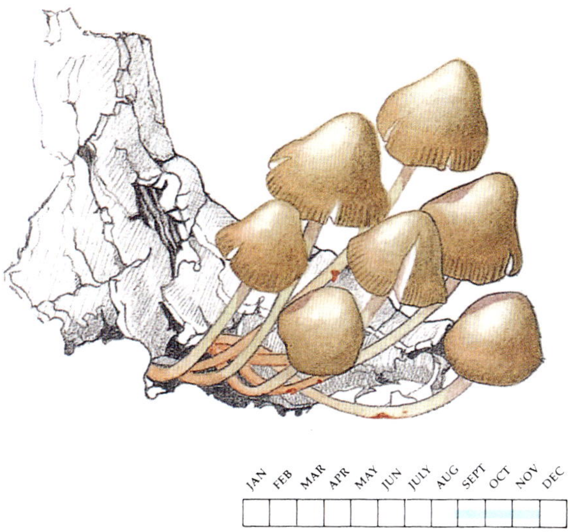

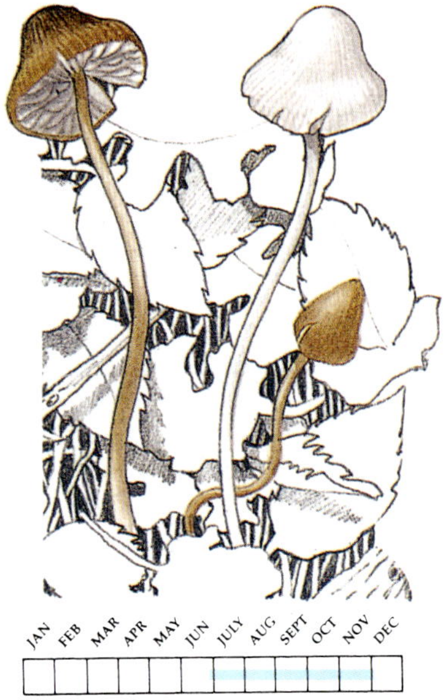

MILK-DROP MYCENA

Mycena galopus

Cap diameter: 1-2cm

Height: up to 10cm

Characteristics: The cap is conical at first but expands and becomes bell-shaped with age. Mature specimens may become rather flattened. The cap colour is pale-brown to buff, darkening tan towards the centre. The broad margin is deeply lined and grooved and the suspended gills may produce a scalloped effect. The gills are greyish-white. Spores white. The stem is long and slender with white, downy hairs covering the base. When broken or damaged, it exudes a white, milky latex, hence the English name. Edible but not worth considering. The flesh is white.

Range and habitat: A common and widespread species throughout the region. Grows in woodlands and hedgerows among leaf-litter.

Similar species: *M. galopus* is extremely variable in colour. In var. *nigra* it is almost black and var. *alba* (=*candida*) it is white. All forms exude white latex.

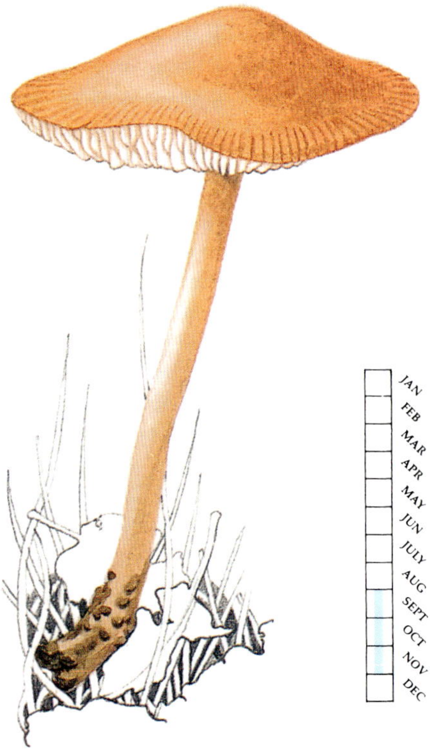

LILAC MYCENA

Mycena pura

Cap diameter: 2-5cm

Height: up to 10cm

Characteristics: The cap is a rounded, conical-shape at first but expands and flattens with age, often with a broad, shallow umbo. The surface is smooth and the margin is lined and grooved when wet. The cap colour varies from pale lilac to pinkish-orange but the margin is generally darker than the cap centre. The gills are whitish-pink. Spores white. The hollow stem is the same colour as the cap. It is rather tough and the base is covered in white, fibrous hairs. Edible but not worth considering. Flesh white, smelling of radish when crushed.

Range and habitat: A common and widespread species in Britain and northern Europe. Grows in mixed and deciduous woodland among leaf-litter and fallen twigs.

Similar species: *M. pelianthina* is similar in size and shape and is greyish-lilac in colour but with dark-bordered gills. It grows in Beech woods and also smells of radish.

MYCENA SANGUINOLENTA

Cap diameter: 1-2cm
Height: up to 8cm
Characteristics: The cap is conical at first but expands and flattens with age with distinct umbo. The cap colour is dull orange or reddish-brown, much darker on and around the central umbo. The margin is striate and splits and tears in mature specimens. The gills are whitish with a dark-red edge. Spores white. The slender stem is the same colour as the cap and exudes a deep red latex when broken. Edible but not worth considering. The flesh is a dull orange colour.
Range and habitat: A common and widespread species in Britain and northern Europe. It grows in a wide variety of habitats from mossy garden lawns to woodland floors and on heathy soils.
Similar species: *M. flavoalba* is a distinctive species which is also commonly found growing on lawns. The flattened conical cap is whitish intensely tinged with yellow towards the centre. The gills and stem are also whitish, tinged with yellow.

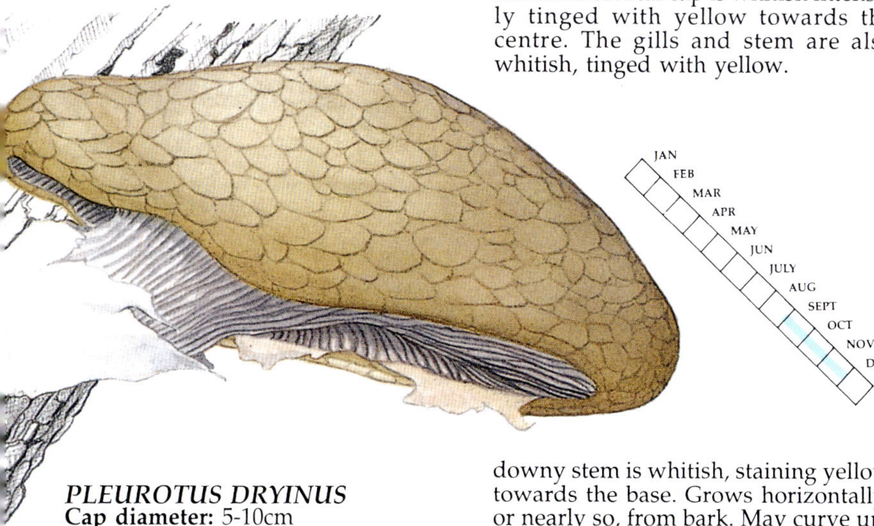

PLEUROTUS DRYINUS

Cap diameter: 5-10cm
Stem length: up to 5cm
Characteristics: The cap is a flattened dome-shape at first but later expands and flattens further with a broad umbo. Mature specimens may have a depressed centre. The cap colour is pale-buff to greyish-brown, darker towards the centre. The surface has a felty coat which soon cracks into woolly scales. The gills are whitish, staining yellow with age, and are strongly decurrent. Spores white. The downy stem is whitish, staining yellow towards the base. Grows horizontally, or nearly so, from bark. May curve upwards slightly. Edible but not worth considering. Flesh white.
Range and habitat: A widespread and locally common species in Britain. It grows on the trunks of Oak and occasionally also on Elm.
Similar species: The unrelated *Lyophyllum ulmarium* grows in clumps on the trunks of Elms and other deciduous trees. The 15cm diameter cap is pale brown and the surface sometimes cracks. Gills are not decurrent.

OYSTER MUSHROOM

Pleurotus ostreatus

Cap diameter: 7-14cm

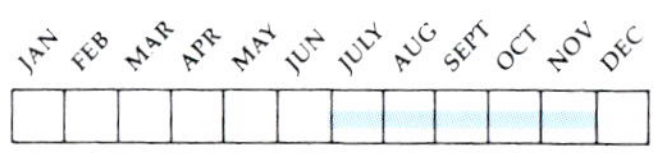

Characteristics: The English name derives from the appearance of the brackets which resemble oyster shells. The smooth cap surface is pale-brown but generally darker brown or blue-grey in younger specimens. The gills are whitish and decurrent and the spore mass is lilac. The stem is short and stout with the brackets arising almost directly from the host tree. Considered edible and good by many people but best eaten young. The flesh is white.

Range and habitat: A widespread and common species in Britain and northern Europe. Found in deciduous woodland and most frequently associated with Beech. It grows in overlapping masses, often low down. Sometimes found on stumps.

Similar species: *P. cornucopiae* has a yellow-buff cap with slightly paler decurrent gills and short stem. It grows in dense clumps with fused stem bases on deciduous trees and is quite widespread.

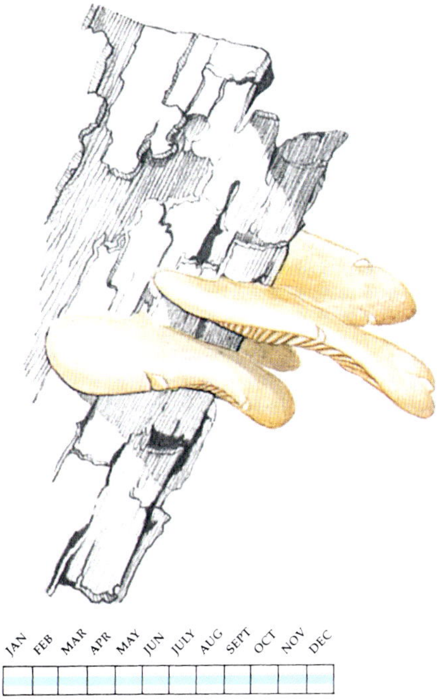

PANELLUS STIPTICUS

Cap diameter: 1-5cm

Characteristics: The flattened cap is semi-circular or kidney-shaped and often has a lobed margin. The surface is slightly downy at first but becomes smooth with age. The cap colour is reddish-buff to pale-brown. The gills are pale brown. Spores white. The stout, whitish stem is very short. It is tapering and grows horizontally from trunks and branches. Not edible. The flesh is yellowish-white and has a bitter taste.

Range and habitat: A fairly common and widespread species in Britain and northern Europe. It grows in overlapping clumps on dead branches and stumps of deciduous trees, especially Oak.

Similar species: *P. serotinus* has a 5-10cm diameter, rounded cap which is yellowish-green at first, becoming orange-brown with age. The gills are yellowish and the stem is short and stout. It grows in overlapping clumps on fallen branches and dead timber and is widespread but rather uncommon throughout the region.

AGROCYBE EREBIA

Cap diameter: 3-6cm

Height: up to 9cm

Characteristics: The cap is domed at first but later expands and flattens with a broad umbo. The surface is slightly sticky in wet weather and the margin is sometimes wavy and irregular in older specimens. The cap colour is orange-brown but it dries paler. The gills are dark brown. Spores brown. The stem is tough and fibrous and brown in colour, darkening towards the base. It bears a pale ring. Edible but not recommended because it can be confused with dangerous species. The flesh is brown.

Range and habitat: Widespread and common in leaf-litter or bare soil.

Similar species: *A. cylindracea* (*=aegerita*) grows in clumps on willow and poplar stumps. The 10cm diameter cap is pale buff, darkening towards the centre. The gills are brown and the stem, which bears a ring, is pale buff. *A. praecox* grows in grassy places and has an orange-yellow cap darkening towards the centre. The gills are brown. Stem and ring yellowish-buff.

CONOCYBE TENERA

Cap diameter: 2-4cm
Height: up to 10cm
Characteristics: The cap is generally conical with a rounded top but it sometimes expands and becomes bell-shaped with age. The cap surface is smooth and the cap colour is buff or orange-brown. In mature specimens, the margin may split or tear. The gills are reddish-brown to dull orange in colour. Spores brown. The stem is long and slender and is sometimes rather undulating. It is pale-brown or buff and rather delicate. Not edible. The flesh is pale-brown.
Range and habitat: A widespread species in Britain and northern Europe and locally common in suitable habitats. Grows along grassy woodland rides and field margins.
Similar species: *C. pseudopilosella* has a flattened, bell-shaped cap which is a rich brown colour, sometimes darkening around the margin. The gills and the long, slender stem are brown. It is rather scarce and grows on garden lawns and along grassy woodland rides.

BOLBITIUS VITELLINUS

Cap diameter: 1-4cm
Height: up to 10cm
Characteristics: The cap is a rounded conical shape at first but soon expands and flattens with age. The surface is bright yellow and sticky when young but fades to pale yellow or buff with age and may look shiny. The cap margin is grooved, blue-grey in colour and sometimes splits with age. The gills are yellow at first but turn reddish with age. Spores brown. The slender stem is yellowish and rather downy towards the base. Not edible. The flesh is whitish.
Range and habitat: A common and widespread species in Britain and northern Europe. Generally found growing on horse dung or well-manured soil but also occurs on wood chips and rotting straw bales.
Similar species: *Conocybe lactea* has a pale yellow cap which is shaped like a bellflower. The gills are reddish and the stem is slender and white. It is rather uncommon and generally grows in grassy places but may be found on sawdust.

HYGROPHORUS HYPOTHEJUS

Cap diameter: 2-6cm
Height: up to 8cm
Characteristics: The cap is a flattened dome-shape at first with an inrolled margin but it later expands and flattens further with age. In mature specimens, the cap often has a depressed centre. The cap surface is rather sticky and the colour is olive-brown to chocolate brown. The decurrent gills are widely spaced and bright orange-yellow in colour. Spores white. The stem is robust and whitish-yellow in colour. It is rather sticky below the ring-like zone. Edible but not recommended. The flesh is whitish but bruises and stains orange.
Range and habitat: Common and widespread throughout the region in suitable habitats. Grows in coniferous woodland among pine needles.
Similar species: *H. dichrous* (=*persoonii*) has a flattened, brown cap with a slight umbo. The surface is slimy. The gills are whitish and the stem is whitish streaked with brown and sticky below the ring-like zone. Grows in deciduous woodland.

HYGROPHORUS CHRYSODON

Cap diameter: 2-6cm
Height: up to 8cm
Characteristics: The cap is domed at first but expands and flattens with age, often with a broad, shallow umbo. The yellowish-white to pale buff cap surface is sticky and slimy. The margin is often patchily covered in yellow scales. The gills are widely spaced and buffish-white, tinged with yellow at the margin. Spores white. The stem is whitish with yellow scales and is rather sticky. Edible but not recommended. The flesh is buffish-white with a strong, mushroomy smell.
Range and habitat: A rather local and uncommon species in Britain and northern Europe. It grows in deciduous woodland, typically Oak or Beech, amongst leaf-litter.
Similar species: Goat Moth Wax-cap, *H. cossus*, is white in all parts and lacks the yellowish scales of *H. chrysodon*. The flesh is white and has a strong and unpleasant smell similar to that of the larva of the Goat Moth. A common and widespread species in Beech woods on chalky soil.

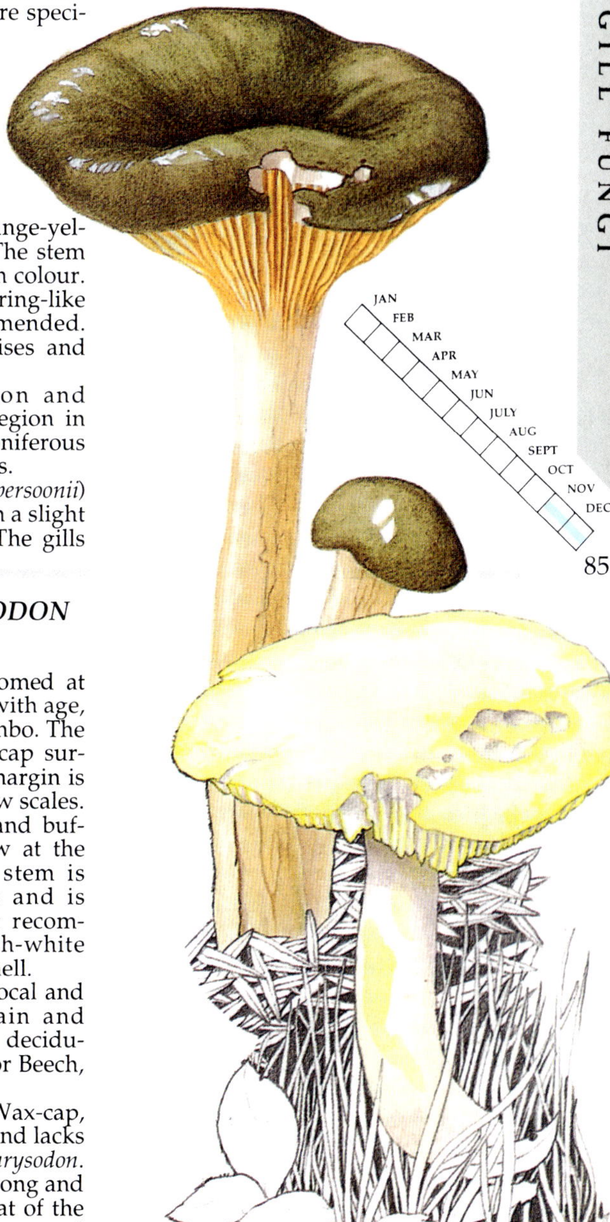

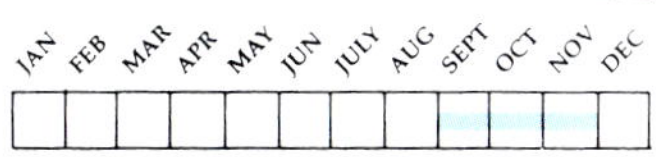

HYGROPHORUS CHRYSASPIS

(=*Discoxanthus chrysaspis*)

Cap diameter: 2-6cm
Height: up to 8cm
Characteristics: The cap is domed at first but soon expands and flattens with age and often bears a slight umbo. The cap is smooth and sticky and the colour is white, sometimes staining yellow with age. The thick, waxy gills are whitish but stain yellow with age. They are slightly decurrent and widely spaced. Spores white. The stem is white and rather sticky and slimy. Large parts of the fungus are eaten by slugs and insects. Not edible. The flesh is whitish and slimy with a strong smell.
Range and habitat: A widespread and locally common species in Britain and northern Europe. It grows in deciduous woodland amongst leaf-litter and is most frequently found under Beech trees.
Similar species: The Ivory Wax-cap, *H. eburneus*, has an ivory-white, slimy cap and stem. The gills are white and decurrent. A local and uncommon species also found in Beech woods.

MEADOW WAX-CAP

Hygrocybe pratensis

Cap diameter: 4-8cm
Height: up to 8cm
Characteristics: At first the cap is conical but it soon expands and flattens, often with a broad, shallow umbo. The cap is waxy and slimy and bright orange or tan in colour. The margin may split and crack with age. The whitish gills are waxy to touch, widely spaced and decurrent. Spores white. The stem is pale yellowish-white and tapers towards the base. Edible and considered good by many people. The flesh is whitish with a mushroomy smell.
Range and habitat: A common and widespread species in Britain and northern Europe. Grows in grassy fields and meadows.
Similar species: *H. intermedia* has a deep, golden-yellow conical cap. The gills are white and the stem is golden-yellow, becoming white towards the base. Edible. Widespread but rather uncommon. Grows in meadows and grassy fields, often on calcareous soils.

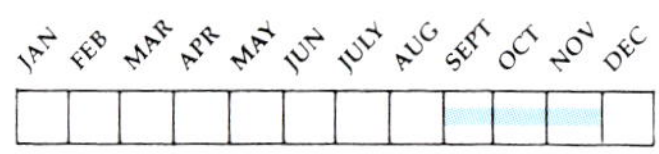

GOLDEN WAX-CAP

Hygrocybe chlorophana
Cap diameter: 3-7cm
Height: up to 8cm
Characteristics: The cap is conical at first but soon expands and flattens with age, often developing a slight umbo. The cap surface is very sticky and lemon yellow or yellow-buff in colour. The waxy gills are widely spaced and yellowish-buff in colour. Spores white. The stem is yellow and is extremely sticky. Edible but not worth considering.
Range and habitat: A common and widespread species in Britain and northern Europe. It grows in fields and meadows and along grassy woodland rides under deciduous trees.
Similar species: The Pink Wax-cap, *H. calyptraeformis*, has a pinkish, conical cap which splits around the margin with age. The gills are widely spaced and whitish and the fragile stem is pink and extremely sticky. Edible but not worth considering. Widespread but not common. Grows in meadows and along grassy woodland rides.

SCARLET WAX CAP

Hygrocybe coccinea
Cap diameter: 2-4cm
Height: up to 6cm
Characteristics: The cap is a rounded, conical shape at first but later expands and becomes rather bell-shaped. Some specimens may be rather irregular in shape. Young specimens are rather sticky and the cap colour is deep scarlet. The gills are widely spaced and are scarlet with an orange-yellow margin. Spores white. The stem is scarlet, becoming paler towards the base, and is rather dry. Edible and considered good by many people. The flesh is reddish.
Range and habitat: A widespread species in Britain and northern Europe and sometimes locally common. Grows in fields and meadows and along grassy woodland rides and margins.
Similar species: *H. konradii* has a conical, orange-yellow cap, often rather irregular in shape. The gills and stem are yellow, the latter becoming white towards the base. An uncommon species which grows in grassland, often on calcareous soils.

CONICAL WAX-CAP

Hygrocybe conica
Cap diameter: 2-5cm
Height: up to 7cm
Characteristics: The cap is conical at first but expands with age, always retaining the acute umbo. Mature specimens may have a distinctly lobed, almost segmented cap. The cap is dry and slightly fibrous. It is orange-yellow at first but blackens when bruised and with age. The gills are pale yellow-orange and waxy and may blacken with age. Spores white. The stem is yellowish and fibrous and may stain black with age or bruising. Edible but not worth considering. The flesh is yellow and bruises black.
Range and habitat: Common and widespread throughout the region. It grows in fields and meadows and sometimes on roadside verges.
Similar species: The Blackening Wax-cap, *H. nigrescens*, grows in fields and along grassy, woodland rides. The cap is orange-red or scarlet, the gills are whitish and the stem yellow. All parts blacken with age or bruising. Edible but not recommended.

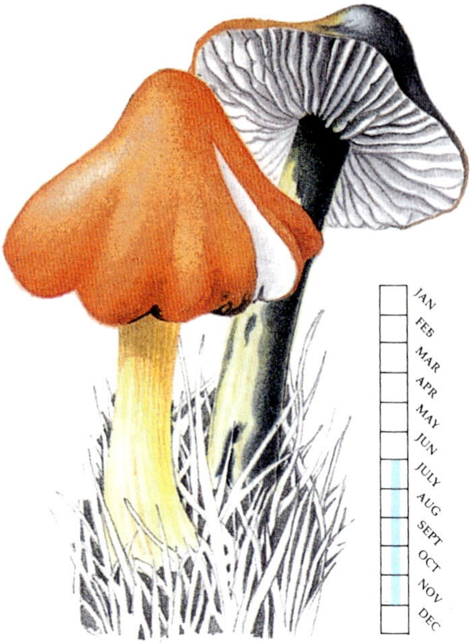

PARROT WAX-CAP

Hygrocybe psittacina
Cap diameter: 1-3cm
Height: up to 5cm
Characteristics: The cap is conical at first but expands with age, becoming bell-shaped or flattened with a distinct umbo. The cap surface is greenish in young specimens and extremely sticky but with age the cap becomes yellowish. The gills are widely separated and greenish-yellow. Spores white. The stem is smooth and slender, yellow in colour becoming greenish towards the apex. It is extremely slimy to touch. Edible but not worth considering because of the slime. The flesh is yellowish-white.
Range and habitat: A widespread species in the region which is sometimes locally common. Grows in grassy habitats on lawns and in fields and meadows.
Similar species: *H. langei* has a flattened, conical cap which is orange or golden-yellow in colour. The stem and gills are also golden-yellow. Rather uncommon. Grows in grassy meadows and fields.

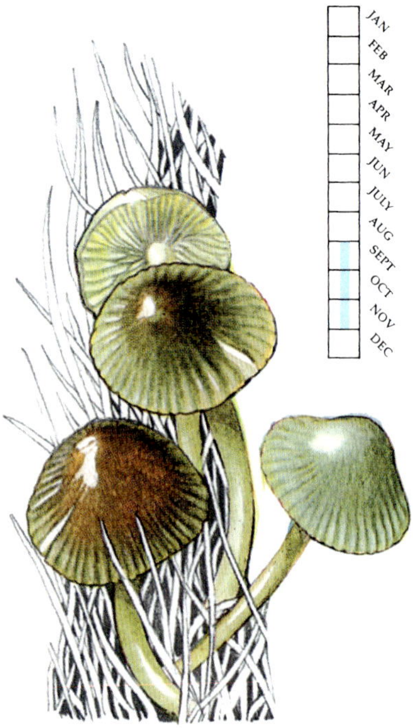

CRIMSON WAX-CAP

Hygrocybe punicea

Cap diameter: 3-6cm
Height: up to 10cm
Characteristics: The cap is conical at first but expands and becomes bell-shaped with age. Often rather irregular in shape in mature specimens with the cap margin lobed and splitting. The cap surface is rather sticky or waxy and cherry-red or crimson in colour. The gills are deep-red with a paler margin. Spores white. The stem is tough and fibrous, often rather distorted in appearance and tapering towards the apex. It is orange-yellow in colour becoming whitish towards the base. Edible and considered good by some people. Flesh whitish, becoming darker in the cap.
Range and habitat: A widespread but distinctly uncommon species throughout the region. It grows in fields, meadows and in grassy woodlands.
Similar species: *H. splendissima* is similar in colour and appearance but less robust. The flesh is yellowish. It grows on acid grassland, becoming more common further north.

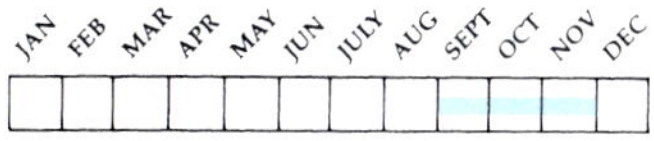

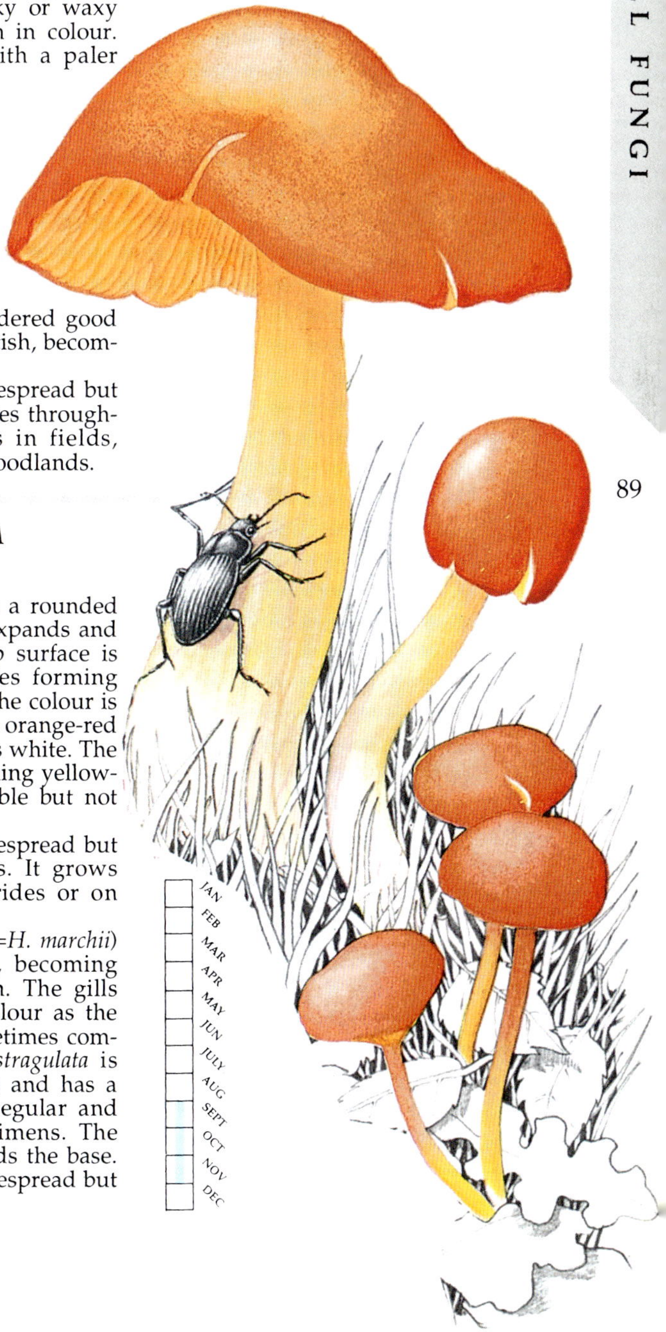

HYGROCYBE MINIATA

Cap diameter: 0.5-1cm
Height: up to 5cm
Characteristics: The cap is a rounded conical shape at first but expands and flattens with age. The cap surface is slightly fibrous, sometimes forming small scales in the centre. the colour is bright scarlet. The gills are orange-red with a paler margin. Spores white. The stem is deep scarlet, becoming yellowish towards the base. Edible but not worth considering.
Range and habitat: A widespread but rather uncommon species. It grows along grassy woodland rides or on heathlands.
Similar species: *H. reidii* (=*H. marchii*) has a golden-yellow cap, becoming paler towards the margin. The gills and stem are the same colour as the cap. Widespread and sometimes common in grassy place. *H. stragulata* is reddish-brown throughout and has a flattened cap which is irregular and often split on older specimens. The stem becomes paler towards the base. Grows on heathlands. Widespread but not common.

BLACKISH-PURPLE RUSSULA

Russula atropurpurea

Cap diameter: 5-10cm
Height: up to 8cm
Characteristics: The cap is domed at first but later expands and flattens with the centre slightly depressed in mature specimens. The cap colour is deep-red or crimson becoming almost black towards the centre. The surface sometimes has paler patches. The gills are creamy white and closely packed. Spores cream. The stem is white and sometimes stained yellowish towards the base. Edible and considered good by some people. The flesh is white and has a fruity smell.
Range and habitat: A common and widespread species in Britain and northern Europe. Grows among leaf-litter in both deciduous and coniferous woodland.
Similar species: *R. alutacea* has a variable cap colour, from brown to purplish-red. The centre is sometimes paler. The stem and gills are yellowish. A rather uncommon species throughout the region which grows in deciduous woodlands.

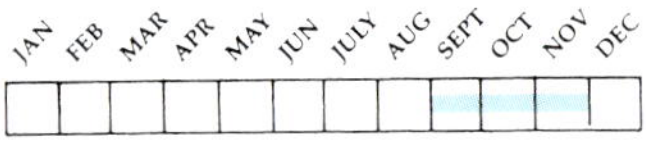

BRIGHT YELLOW RUSSULA

Russula claroflava

Cap diameter: 5-10cm
Height: up to 10cm
Characteristics: The cap is domed at first but later expands and flattens. The surface is sticky and shiny in young specimens but dries with age. The cap colour is bright yellow but may blacken with age. The gills are yellowish. Spores cream. The white stem is slightly tapering. Edible and considered good by many people. However, it may be confused with less palatable species of *Russula*. The flesh is white and soft.
Range and habitat: A common and widespread species in Britain and northern Europe in suitable habitats. Grows in Birch woodlands, especially where the soil is damp and boggy.
Similar species: *R. aeruginea* has a flattened cap which is greyish-buff in colour, sometimes with a greenish tinge. The stem is whitish and the gills are creamy-white. Grows under Birch and in coniferous woodland. A common and widespread species throughout the region.

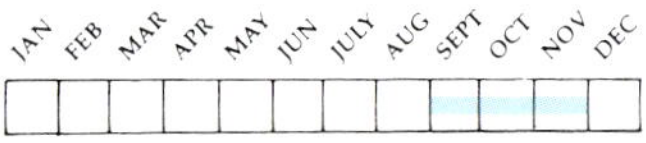

SICKENER

Russula emetica

Cap diameter: 5-10cm

Height: up to 10cm

Characteristics: The cap is rounded at first but later expands and flattens often with a depressed centre. The cap surface is shiny and sticky when wet and the skin peels easily. Mature specimens have a grooved margin. The colour is bright red or crimson. The gills are creamy white and the spores are white. The stem is white and has a slightly bulbous base. Poisonous, as the English name suggests. The flesh is white but pink beneath the cap skin.

Range and habitat: A common and widespread species throughout the region. Grows in coniferous woodland.

Similar species: The Beechwood Sickener is similar in appearance with a 4-9cm diameter cap and grows under Beech. It is also poisonous and common and widespread. *R. betularum* has a pale pink cap and white gills and stem. It is widespread and common, growing under beech, and is a poisonous species.

COMMON YELLOW RUSSULA

Russula ochroleuca

Cap diameter: 5-10cm

Height: up to 10cm

Characteristics: The cap is domed at first but later expands and flattens with age. Mature specimens may have the cap centre slightly depressed and the margin furrowed. The cap colour is bright yellow to orange yellow although the skin may peel. The gills are creamy-white. Spores off-white. The yellowish-white stem sometimes tapers and may be hollow. Edible although not to everyone's taste. The flesh may have a hot taste.

Range and habitat: An extremely common and widespread species in Britain and northern Europe. Grows in wooded habitats, under both coniferous and deciduous trees.

Similar species: *R. farnipes* has a buff-yellow cap, 2-6 cm in diameter, and with a furrowed margin and sometimes a depressed centre. The stem, gills and flesh are creamy white. A common and widespread species throughout the region, growing in deciduous woodland.

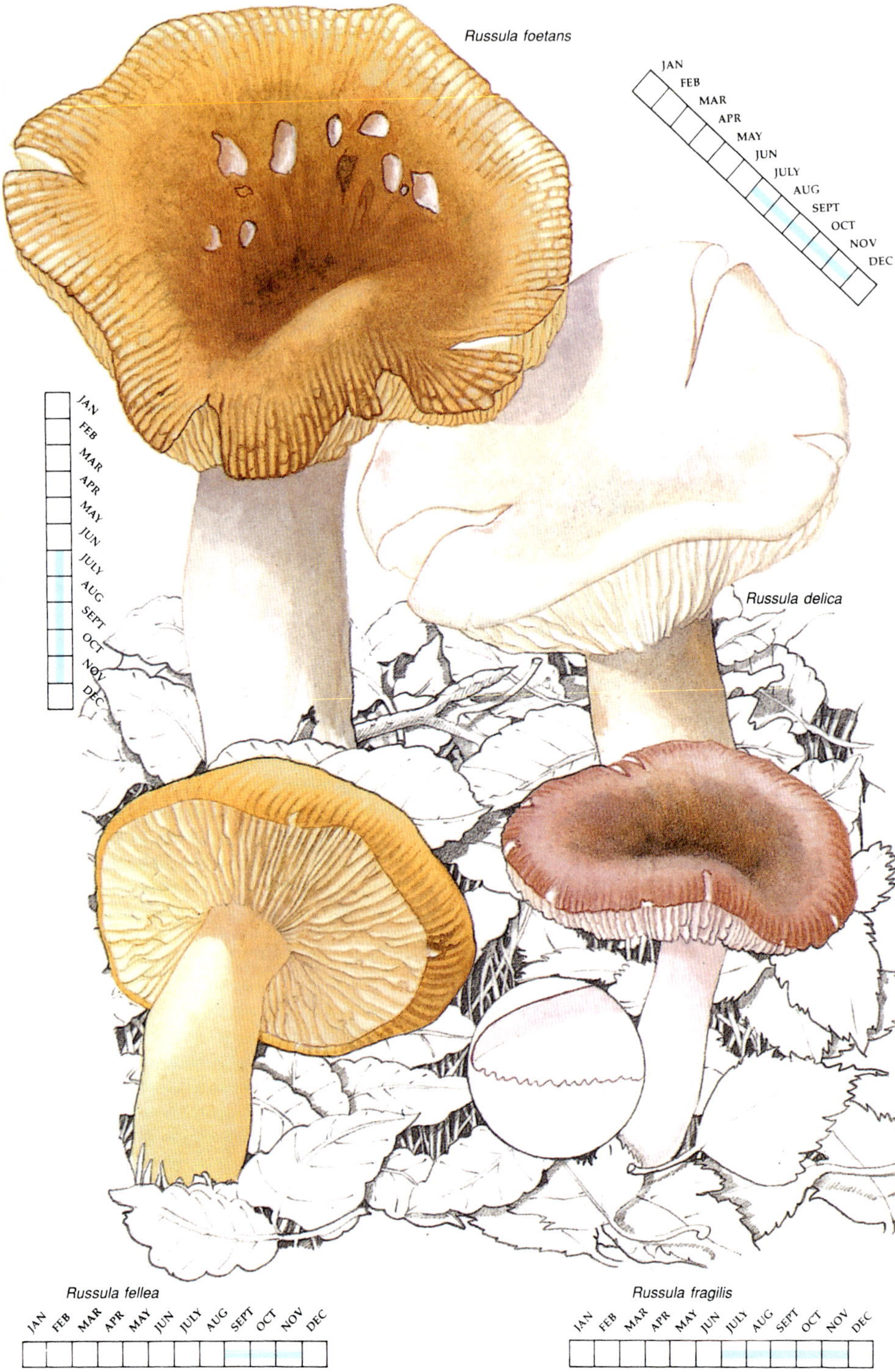
Russula foetans
JAN FEB MAR APR MAY JUN JULY AUG SEPT OCT NOV DEC
JAN FEB MAR APR MAY JUN JULY AUG SEPT OCT NOV DEC
Russula delica
Russula fellea
JAN FEB MAR APR MAY JUN JULY AUG SEPT OCT NOV DEC
Russula fragilis
JAN FEB MAR APR MAY JUN JULY AUG SEPT OCT NOV DEC

STINKING RUSSULA

Russula foetens

Cap diameter: 6-12cm
Height: up to 12cm
Characteristics: The cap is domed at first but later expands and flattens with age. In mature specimens, the cap centre may be depressed. The cap colour is a rich, orange-buff, becoming darker tan towards the centre. The margin is distinctly furrowed and grooved with small scales and the surface is sticky when wet. The gills are creamy buff in colour and the stem is white and swollen. Spores cream. Not edible. As the English name suggests the white flesh smells rancid and has a bitter taste.
Range and habitat: Common and widespread throughout the region. It grows in deciduous and coniferous woodland, especially in damp soil.
Similar species: *R. laurocerasi* is similar but has a dull buff-tan cap and creamy-white gills and stem. It is a common and widespread species growing especially in Oak woodland and has a strong smell of bitter almonds or marzipan. Not edible.

MILK-WHITE RUSSULA

Russula delica

Cap diameter: 5-15cm
Height: up to 10cm
Characteristics: The cap is domed at first but later expands and flattens. Mature specimens may become funnel-shaped with an inrolled margin. The cap colour is whitish but it is sometimes irregularly blotched with brown. The gills are decurrent and white. Spores white. The stout, short stem is white and tough. Edible but not worth considering. The flesh is white and has a hot and bitter taste.
Range and habitat: A common and widespread species in Britain and northern Europe. It grows in wooded habitats under both coniferous and deciduous trees.
Similar species: *R. densifolia* is similar in appearance. It is common and widespread in deciduous woodland, especially under Oak, as well as on heathlands. The funnel-shaped cap is clay-brown, sometimes darker towards the centre, and the gills and stem are buffish-white. Edible but not worth considering.

GERANIUM-SCENTED RUSSULA

Russula fellea

Cap diameter: 5-10cm
Height: up to 10cm
Characteristics: The cap is domed at first but later expands and flattens with age. In mature specimens, the centre becomes depressed with a broad umbo. The cap surface is slightly sticky when wet and the margin is furrowed. The colour is orange-buff or straw-coloured. The gills are creamy-white. Spores cream. The stem is short and fragile and yellowish-buff in colour. Edible but not recommended. The flesh is white and has a hot taste and strong smell of Geraniums.
Range and habitat: A common and widespread species. It grows in deciduous woodland, especially under Beech trees.
Similar species: *R. amoenolens* (=*R. sororia*) has a grey-brown, 3-6cm diameter cap and whitish stem and gills. The flesh is white with a strong smell of rancid cheese. A rather uncommon species growing in deciduous woodland, especially under Oak trees.

TOOTHED-GILL RUSSULA; FRAGILE RUSSULA

Russula fragilis

Cap diameter: 3-6cm
Height: up to 6cm
Characteristics: The cap is rounded at first but later expands and flattens with the centre slightly depressed. The margin is furrowed and the cap flesh is extremely brittle. The colour is usually reddish-violet darkening to almost black towards the centre. The gills are white and have a minutely serrated edge. Spores white. The stem is white and extremely fragile. May be edible but best avoided. Flesh white with a fruity smell and hot taste.
Range and habitat: A common and widespread species throughout the region. It grows in wooded habitats under coniferous and deciduous trees.
Similar species: *R. aquosa* has reddish-violet cap, 4-8cm in diameter, which is rather flattened with a depressed centre. The gills and stem are white and the flesh has a hot taste and slight smell of coconut. Grows in damp woodland. Rather uncommon throughout the region.

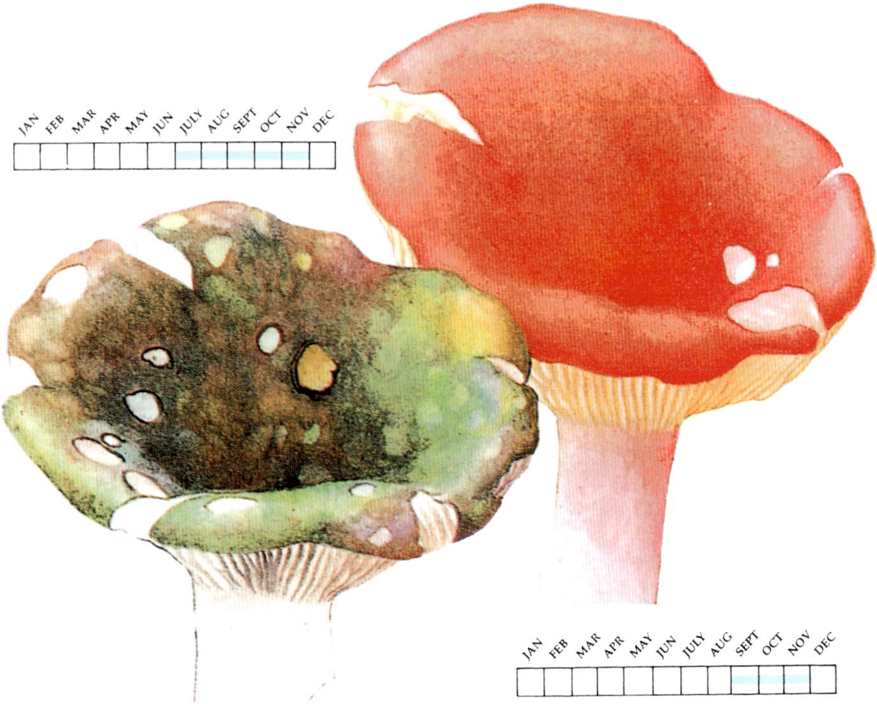

THE CHARCOAL BURNER

Russula cyanoxantha
Cap diameter: 5-15cm
Height: up to 12cm
Characteristics: The cap is domed at first but later expands and flattens with age, sometimes with a depressed centre. The cap is firm-fleshed and slightly slimy in wet weather. The colour is usually a varied patchwork of violet, green, flesh-pink and grey-brown but may occasionally be a single, uniform colour. The gills are white and slightly decurrent and are not brittle. Spores white. The stem is white, sometimes with a slightly bulbous base. Edible and considered good by many people. Flesh white.
Range and habitat: An extremely common and widespread species in suitable habitats. Grows amongst leaf-litter in deciduous woodland.
Similar species: *R. ionochlora* has a cap mottled with greenish-grey, reddish-brown and buff. The margin is generally a uniform grey-brown flushed with red. The gills and stem are creamy white. Rather uncommon. Grows in Beech woodland.

RUSSULA LEPIDA

Cap diameter: 5-10cm
Height: up to 10cm
Characteristics: At first the cap is domed but later it flattens and expands, sometimes with a depressed centre. The cap margin is furrowed and the skin seldom peels. The cap colour is scarlet to deep pink and is sometimes dusted with white powder. The gills are creamy-white and brittle. Spores white. The stout, tough stem is white sometimes flushing pink towards the swollen or bulbous base. Not edible. Flesh white with a taste of cedarwood and fruity smell.
Range and habitat: A common and widespread species throughout the region. It grows in leaf litter in deciduous woodland, especially under Beech trees.
Similar species: *R. nitida* has a dull-red, flattened cap with a depressed centre. The gills are creamy-white and the stem is white. It is common and widespread and grows under Birch trees, especially with damp Sphagnum moss. Edible. Some specimens have a hot taste.

RUSSULA LUTEA

Cap diameter: 2-5cm
Height: up to 6cm
Characteristics: The cap is an irregular domed-shape at first but later expands and flattens. The margin is generally furrowed and the centre may be depressed in some mature specimens. The cap colour is golden-yellow, tinged with flesh-pink in some specimens. The skin often peels. The gills are orange-yellow and the stout stem is white. Spores orange. Edible but not worth considering because of its soft texture. The flesh is white with a mild taste and a slight smell of apricots in fresh specimens.
Range and habitat: A common and widespread species in Britain and northern Europe. It grows in deciduous woodland, especially under Oaks, amongst leaf-litter.
Similar species: *R. puellaris* has a reddish, 3-6cm diameter cap with a depressed centre. The gills and stem are whitish at first but stain and bruise yellow with age. A common species which grows in deciduous woodland.

RUSSULA QUELETII

Cap diameter: 5-10cm
Height: up to 10cm
Characteristics: The cap is domed and rounded at first but soon expands and flattens with a slight central depression. The cap colour is deep wine-red or reddish-violet but this fades with age. The gills are slightly decurrent and creamy-buff in colour. Spores cream. The stem generally has a slightly bulbous base and is flushed with red and streaked and veined darker. Not edible. The flesh is white and has a bitter taste and fruity smell.
Range and habitat: A rather local and uncommon species throughout the region, often seeming to prefer upland areas. Grows among pine needles in coniferous woodland including plantations.
Similar species: *R. sanguinea* is similar in appearance but has a brighter red cap, 6-12cm in diameter. The gills are creamy-yellow and the stem is stout and flushed with red. It grows in coniferous woodland and is locally common.

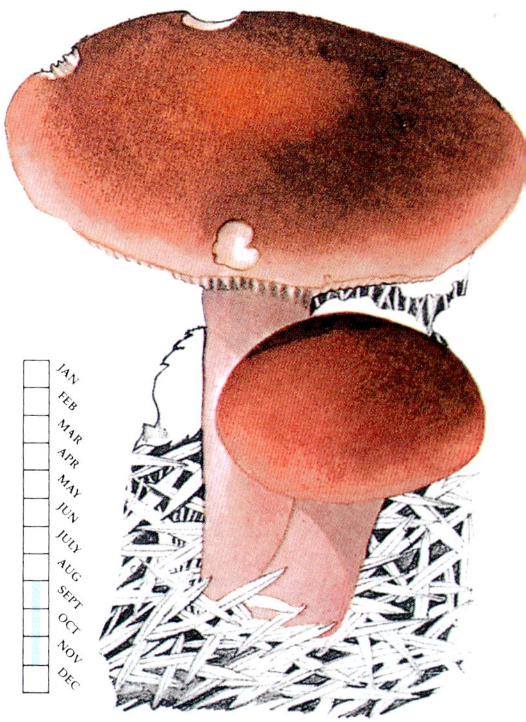

BARE-EDGED RUSSULA

Russula vesca

Cap diameter: 5-10cm
Height: up to 10cm
Characteristics: The cap is distinctly domed at first but soon expands and flattens with age. The cap colour is variable, from reddish-brown to olive-brown, and is usually rather mottled. The skin around the margin retracts leaving the white, ribbed flesh exposed beneath, hence the English name. The gills are creamy-white. Spores white. The stout stem is white with a rather pointed base. Edible and considered good by some people. The flesh is white.
Range and habitat: A common and widespread species throughout the region. Grows in leaf-litter in deciduous woodlands.
Similar species: *R. grisea* has a grey-brown cap, mottled with pink especially towards the centre. The cap flesh turns pale lilac on exposure to air. The stem and gills are whitish. Grows in deciduous woodland, especially under Beech, and is rather rare and local.

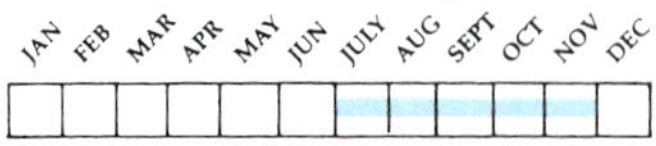

RUSSULA XERAMPELINA

Cap diameter: 5-15cm
Height: up to 12cm
Characteristics: The cap is domed at first but later expands and flattens with a shallow, central depression. The cap colour is extremely variable from olive-brown to yellowish-buff to red. It is usually a blotched and mottled mixture of several colours. The gills are creamy-buff and the stout stem is whitish, sometimes stained or streaked darker. Spores ochre. Edible. The flesh is white, has a distinct fishy smell. Turns bright green when treated with Iron sulphate.
Range and habitat: A common and widespread species in Britain and northern Europe. It grows in deciduous woodland, especially under Oak and Beech trees.
Similar species: *R. pulchella* (=*R. exalbicans*) has a flattened buff or pink cap with a shallow depression. The gills are creamy-buff and the white stem sometimes has a bulbous base. It is a common and widespread species, growing in coniferous woodland.

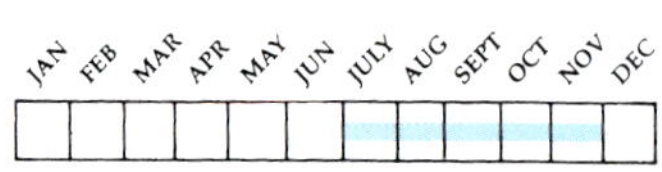

BLACKENING RUSSULA

Russula nigricans

Cap diameter: 5-20cm

Height: up to 12cm

Characteristics: The cap is domed at first but soon expands and flattens with a deep, central depression. The cap is tough and firm to touch. The cap colour is white at first but soon turns brown and finally black as though charred. The gills are buffish-brown. Spores white. The tough stem is a dirty white colour. Edible but not worth considering. The flesh is white at first but turns red and then black with exposure to air. It has a fruity smell.

Range and habitat: An extremely common and widespread species throughout the region. It grows in leaf litter under both deciduous and coniferous trees.

Similar species: *R. albonigra* is less frequent and grows in a variety of wooded habitats. The cap soon becomes black and the gills and stem are greyish-white. The flesh turns directly from white to black on exposure to air without a red intermediate stage.

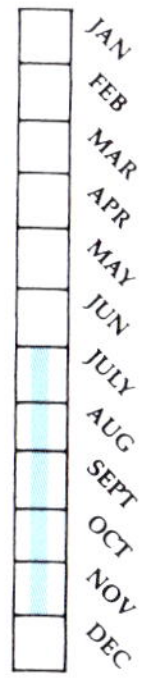

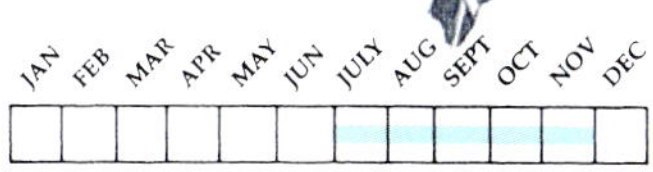

GREEN CRACKING RUSSULA

Russula virescens

Cap diameter: 5-10cm

Height: up to 10cm

Characteristics: The cap is domed and rounded at first but later expands and flattens with an irregular margin and depressed centre. The cap colour is buff to vedigris-green and the surface is covered in darker green scales. The gills are white and brittle. Spores white. The stout, white stem often has a pointed base. Edible and considered good by many people. The flesh is white and has a mild taste.

Range and habitat: A widespread species which is sometimes locally common in suitable habitats throughout Britain and northern Europe. It grows in deciduous woodland, especially under Beech.

Similar species: *R. heterophylla* has flattened 5-10cm diameter cap with a depressed centre. The cap colour is greenish-brown to yellowish-buff and the stem and decurrent gills are whitish. A rather uncommon species throughout the region, which grows in deciduous woodland.

SLIMY MILK-CAP

Lactarius blennius

Cap diameter: 5-10cm
Height: up to 8cm
Characteristics: At first the cap is a flattened dome with a dimpled centre and inrolled margin. This expands and flattens further with age with a deeply depressed centre and margin still inrolled. The cap surface is extremely slimy. It is olive-brown in colour, often blotched with red or dark brown marks which may form concentric rings. The gills are slightly decurrent and dirty white in colour. Spores cream. The stem is pale brown and stout. Edible but not worth considering. The flesh is white and the milk is white, turning grey with a hot taste.
Range and habitat: A common and widespread species throughout the region. It grows in deciduous woodland, especially under Beech.
Similar species: *L. uvidus* is orange-buff in colour, bruising lilac. The milk is whitish. Rather uncommon but more frequent in Scotland growing under Birch.

YELLOW MILK-CAP

Lactarius chrysorrheus

Cap diameter: 4-8cm
Height: up to 8cm
Characteristics: The cap is a flattened dome with an inrolled margin at first but later expands and flattens further with a distinctly depressed centre. Young specimens are usually sticky. The cap colour is pinkish-orange often with darker concentric rings or darker blotches. The decurrent gills are creamy-buff in colour. Spores white. The stout stem is the same colour as the cap. Not edible and possibly poisonous. The flesh is white but stains yellow when cut due to the yellowing milk.
Range and habitat: A common and widespread species throughout the region. It grows in deciduous woodlands, almost exclusively under Oak.
Similar species: *L. helvus* has a flattened, pinkish-buff cap with a slightly depressed centre. The surface is slightly scaly and the milk is clear. Rather scarce, growing in coniferous woodland.

OAK MILK-CAP; OILY MILK-CAP

Lactarius quietus

Cap diameter: 4-8cm
Height: up to 8cm
Characteristics: The cap is domed at first but later expands and flattens with a depressed centre and margin often slightly inrolled. The surface is smooth and not sticky and the colour is reddish-brown to buff, with concentric rings of darker blotches. The gills are decurrent and orange-brown in colour. Spores cream. The stem is the same colour as the cap. Edible but not worth considering. The flesh is off-white and the milk is white with a rancid, oily smell.
Range and habitat: An extremely common and widespread species throughout the region in suitable habitats. It grows exclusively under Oaks.
Similar species: *L. pyrogalus* has a yellowish cap sometimes with concentric rings of darker, orange blotches. The gills are yellow and the stem and flesh are whitish. The milk is white and has a hot taste. A common and widespread species which grows under Hazel.

RED MILK-CAP

Lactarius rufus

Cap diameter: 4-8cm
Height: up to 8cm
Characteristics: The cap is flattened at first with a slightly inrolled margin. Later it expands and acquires a distinctly depressed centre with a prominent, pointed umbo. The cap surface is dry and the colour is brick-red to orange-tan. The gills are slightly decurrent and creamy-buff in colour. Spores cream. The stem is orange-buff in colour. Not edible. The flesh is white and the milk is white and mild at first with a hot after-taste.
Range and habitat: An extremely common and widespread species throughout the region. It usually grows in coniferous woodland but may be found under Birch.
Similar species: *L. subdulcis* is common in deciduous woodland, especially under Beech. The orange-buff cap has a depressed centre, sometimes with a slight umbo and inrolled margin. The gills are whitish and the stem is buff. The milk is white with a mild taste becoming bitter later.

Lactarius blennius

Lactarius chrysorrheus

Lactarius quietus

Lactarius rufus

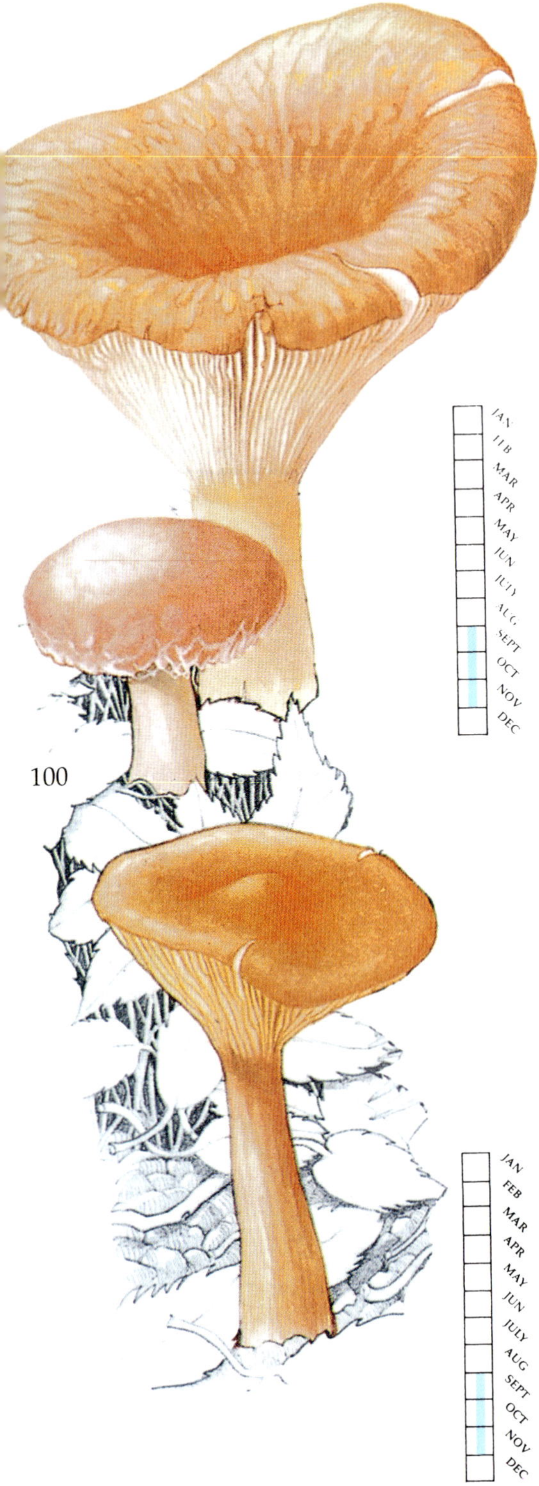

WOOLLY MILK-CAP

Lactarius torminosus

Cap diameter: 4-12cm
Height: up to 10cm
Characteristics: At first the cap is a rounded dome with a markedly in-rolled margin. Later it expands and flattens becoming slightly funnel shaped with a deeply depressed centre. The cap surface is covered in shaggy, woolly fibres. The colour is pinkish-orange, generally with concentric rings of darker orange blotches. The gills are slightly decurrent and flesh-pink. Spores cream. The pinkish-orange stem has downy hairs and is sometimes hollow. May be poisonous. Flesh whitish. Milk white and hot.
Range and habitat: A widespread species which is sometimes locally abundant in suitable habitats. Grows in association with Birch trees and is found in woodlands and on heaths.
Similar species: *L. pubescens* has creamy-white cap, often stained and blotched with yellow. The centre is strongly depressed and the margin is inrolled and woolly. Rather uncommon. Associated with Birch trees.

LACTARIUS TABIDUS

Cap diameter: 2-4cm
Height: up to 6cm
Characteristics: The cap is slightly domed at first but soon expands and flattens with age. Mature specimens have a central depression to the cap often with a tiny umbo. The cap surface is generally smooth although the centre may be slightly wrinkled and it is not sticky. The cap colour is orange-tan to buff-brown. The gills are slightly decurrent and yellow-buff in colour. Spores cream. The brittle stem tapers towards the apex and is the same colour as the cap. Not edible. The flesh is white and the milk is white but turns yellow on a handkerchief.
Range and habitat: A common and widespread species throughout the region. It grows in leaf litter in deciduous woods, often under Birch.
Similar species: *L. hepaticus* grows under conifers in the south. The cap is flattened with a slightly depressed centre and the colour is muddy brown throughout. The milk is white, turning yellow on a handkerchief.

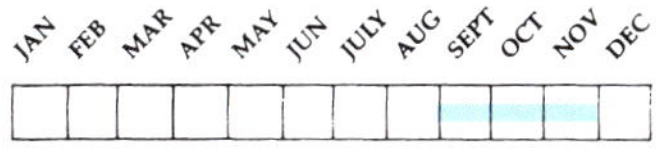

SAFFRON MILK-CAP

Lactarius deliciosus

Cap diameter: 4-10cm

Height: up to 8cm

Characteristics: At first the cap is a flattened dome with a depressed centre and an inrolled margin. It later expands and becomes slightly funnel-shaped. The cap surface is slightly sticky and the colour is saffron-yellow with concentric rings of darker orange blotches. The gills are decurrent and orange-yellow. Spores cream. The stem is orange-buff in colour and often has darker orange pits and blotches. The whole fungus may discolour green in places. Edible and considered good by many people. The flesh and milk are yellowish-orange, eventually turning greenish.

Range and habitat: A widespread species throughout the region but not common in Britain. It grows in coniferous woodland.

Similar species: *L. deterrimus* is very similar but grows under spruce and more readily turns green. The flesh and milk are orange but eventually turn deep red.

GREY MILK-CAP

Lactarius vietus

Cap diameter: 5-10cm

Height: up to 10cm

Characteristics: At first the cap is a flattened dome but this later expands and becomes funnel shaped with age, sometimes with a slight umbo. The cap surface is slightly sticky in damp weather and the colour is greyish-lilac to pale buff. The gills are slightly decurrent and buff to grey-brown. Spores cream. The stem is grey brown to lilac in colour and rather soft. Not edible. The flesh is pale buff and the milk is white at first, turning grey later with a hot taste.

Range and habitat: A common and widespread species throughout the region. It grows in leaf-litter in damp, deciduous woodland, especially under Birch.

Similar species: *L. flavidus* is a rather uncommon species which grows in Beech woods. The cap is 4-6cm in diameter and the fungus is up to 6cm in height. It is pale brown throughout but bruises violet after fifteen minutes. The milk is white and hot.

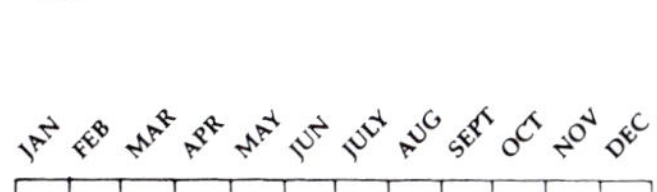

PEPPERY MILK-CAP

Lactarius piperatus
Cap diameter: 5-15cm
Height: up to 10cm
Characteristics: At first the cap is a flattened dome with a depressed centre and inrolled margin. With age the cap becomes funnel-shaped. The cap colour is creamy white and the surface is often covered in darker flaky scales. The gills are slightly decurrent and are yellowish to orange in colour. Spores white. The stem is white and tapers towards the base. Edible but best avoided. Flesh white. The milk is white with a hot, peppery taste.
Range and habitat: A common and widespread species throughout the region. It grows in leaf-litter in deciduous woodland.
Similar species: The Fleecy Milk-cap, *L. vellereus*, is common and widespread, growing in leaf-litter in deciduous woodland. The cap is 10-20cm in diameter and creamy-buff in colour covered with fleecy hairs. The gills are creamy-yellow and decurrent and the stem is the same colour as the cap. The milk is white.

JAN	FEB	MAR	APR	MAY	JUN	JULY	AUG	SEPT	OCT	NOV	DEC

LACTARIUS VOLEMUS

Cap diameter: 5-10cm
Height: up to 12cm
Charateristics: The cap is domed and rounded at first. It expands and flattens with age developing a depressed centre and a slight umbo. The cap surface is dry and is reddish-orange to tawny-brown in colour. The surface may split and crack with age. The gills are decurrent and creamy yellow in colour, bruising darker. The stem is stout and tough and the same colour as the cape. Edible but not recommended. The flesh is whitish with a distinct smell of herrings. The milk is white but darkens slightly on exposure to air.
Range and habitat: A distinctly local species which is generally rather uncommon. It occurs among leaf litter in deciduous woodlands, especially under oak. Can be difficult to spot on the woodland floor.
Similar species: *L. tabidus* is a similar colour but smaller. It is much more common than *L. volemus* and has milk which turns yellow on a handkerchief.

JAN	FEB	MAR	APR	MAY	JUN	JULY	AUG	SEPT	OCT	NOV	DEC

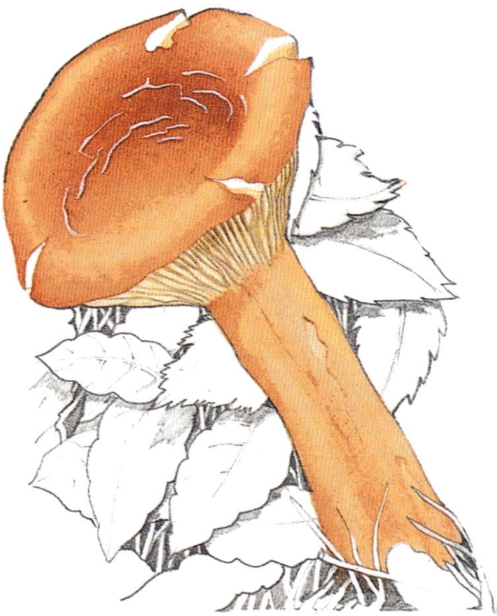

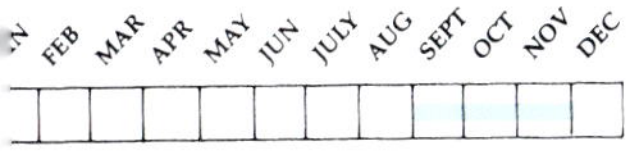

UGLY MILK-CAP

Lactarius turpis

Cap diameter: 5-15cm
Height: up to 12cm
Characteristics: The cap is a flattened dome at first with a depressed centre and inrolled margin. It expands and flattens with age and retains the depressed centre. The cap is rather sticky in damp weather and woolly around the margin. The cap colour is olive-green to brown, sometimes with darker concentric rings. The gills are slightly decurrent and yellowish-brown in colour. Spores cream. The stem is the same colour as the cap. Sticky and often pitted with holes. Not edible. Flesh white. Milk white and hot.
Range and habitat: An extremely common and widespread species. It grows in damp, deciduous woodland, often in association with Birch.
Similar species: *L. fluens* has a flattened cap with a depressed centre and inrolled margin. The colour is buff-brown to straw coloured. The milk is whitish. A rather uncommon species growing under Hornbeam and Beech.

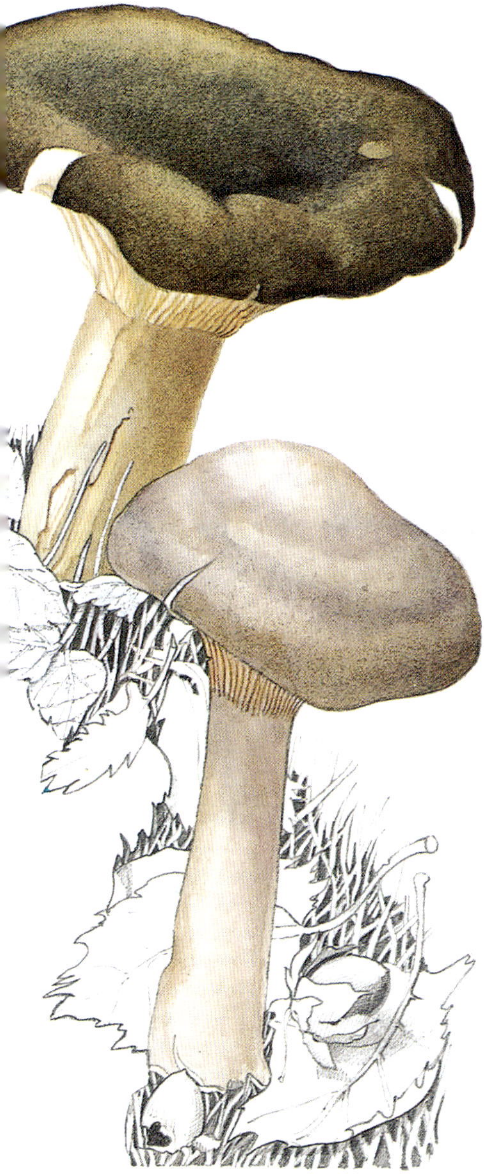

COCONUT MILK-CAP

Lactarius glyciosmus

Cap diameter: 3-6cm
Height: up to 7cm
Characteristics: At first the cap is a flattened dome-shape but this later expands and flattens with age sometimes with a slight umbo. The cap surface is not sticky and the colour is greyish-lilac to buff. The gills are decurrent and yellowish-brown. Spores cream. The stem is pale greyish-buff and rather soft. Edible and considered good by some people. The flesh is yellowish-white and smells of coconut. The milk is white with a hot after-taste.
Range and habitat: Common and widespread. It grows in leaf-litter in deciduous woodland, especially in association with Birch trees.
Similar species: *L. fuliginosus* has a flattened cap, 4-8cm in diameter, with a depressed centre and inrolled margin. The cap colour is dark grey-brown while the stem and gills are yellowish-buff. A rather uncommon species which grows in deciduous woodland, especially under Beech.

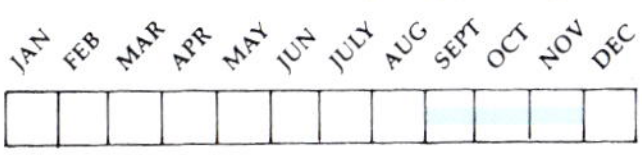

THE BOLETES – FUNGI WITH PORES

This large and wide-ranging group of fungi are characterised by having the spores produced in a network of tubes under the cap rather than on gills. The tube openings are visible as pores beneath the cap and the tube mass is usually easily separable from the cap flesh. The caps are usually very broad and dry rather than sticky. The stems are generally fat and bulbous and may be covered in a network of darker coloured veins.

Almost all species are found on woodland floors growing among leaf-litter or along grassy rides. Although at one time most were included in the genus *Boletus*, many are now included in separate genera. Those covered in this book include *Leccinum, Suillus, Strobilomyces, Gyroporus, Tylopilus* and *Porphyrellus* as well as *Boletus*.

With the exception of some species with red or orange coloured pores, most species are edible. One in particular, the Cep or Penny Bun, *Boletus edulis*, is exceptionally good and is an essential part of French cuisine.

CEP; PENNY BUN

Boletus edulis

Cap diameter: 5-20cm

Height: up to 20cm

Characteristics: The cap is hemispherical at first but later expands and becomes rounded and flattened with age. The cap surface is slightly slippery in wet weather. The colour is rich chestnut-brown, paler around the edge and often with a narrow white margin. The pores are white but may stain yellow with age. Pores olive-brown. The stem is thick and swollen. It is white in colour, often stained and netted with darker brown. Edible and delicious. Considered by many to be the best edible species in the region. Flesh white. Does not stain darker.

Range and habitat: A common and widespread species. It grows mainly in deciduous woodland but also under conifers.

Similar species: *B. aereus* also grows in deciduous woodland but has a dark brown cap, whitish pores and a reddish-brown stem. It is edible and delicious.

SATAN'S BOLETE

Boletus satanus

Cap diameter: 10-25cm

Height: up to 10cm

Characteristics: The cap is rounded at first but later expands and becomes a flattened dome. The cap surface is smooth and the margin is sometimes wrinkled. The colour is whitish although the margin may be flushed with red. The pores are reddish-orange but bruise greenish-blue. Spores olive-brown. The stem is extremely stout and swollen towards the base. It is yellowish in colour staining red towards the base and covered in a rich network of deep red veins. Not edible and may be poisonous. Flesh yellowish becoming pale blue on exposure to air.

Range and habitat: A distinctly uncommon species. It grows in deciduous woodlands and often under Beech or Oak on chalky soil.

Similar species: *B. luridus* has a buff-tan cap, orange pores and a yellow stem staining orange towards the base. It grows in deciduous woodland, especially under Oak and Beech, and is rather uncommon.

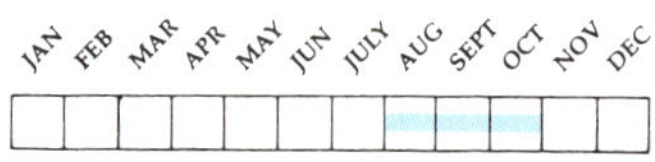

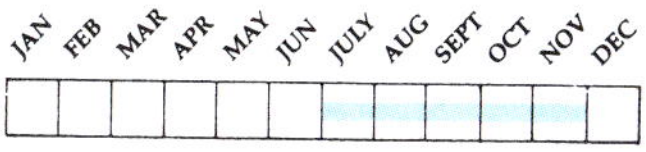

Boletus edulis

BAY-CAPPED BOLETE

Boletus badius

Cap diameter: 5-15cm
Height: up to 15cm
Characteristics: The cap is domed at first but soon expands and flattens slightly. The cap surface is downy at first but later becomes smooth with age. It is slightly sticky in damp weather. The colour is chestnut-brown to tan. The pores are creamy-yellow but bruise blue-green. Spores olive-brown. The stem is straight and pale-brown, netted with darker brown veins. Edible and good. Flesh pale yellow, turning blue on exposure to air.
Range and habitat: A common and widespread species throughout the region. It grows in both coniferous and deciduous woodland.
Similar species: *B. parasiticus* is an unusual species which is yellow-brown throughout. The cap is 2-4cm in diameter and the fungus is up to 5cm tall. It is a parasitic species, growing on the Common Earthball, *Scleroderma citrinum*. It is rather uncommon and is found in deciduous woodland.

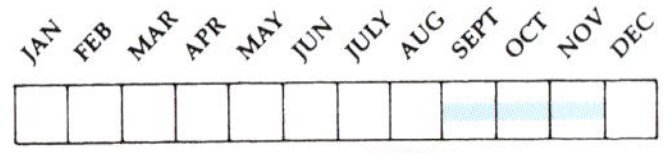

BOLETUS ERYTHROPUS

Cap diameter: 5-15cm
Height: up to 15cm
Characteristics: At first the cap is hemispherical but later it expands and flattens. The cap surface is slightly downy in young specimens but later becomes smooth and slightly sticky. The cap colour is chestnut-brown to tan sometimes becoming paler around the margin. The pores are reddish-orange but bruise blue. Spores olive-brown. The stem is fat and swollen. It is orange-yellow in colour, blotched with red, and lacks a network of veins. Edible but not to be recommended. Flesh yellow but quickly staining blue on exposure to air.
Range and habitat: A common and widespread species throughout the region. It grows in both deciduous and coniferous woodland.
Similar species: *B. purpureus* has a buffish-yellow cap, which stains blue with bruising, red pores and an orange stem with a network of red veins. The flesh is yellowish but rapidly stains blue on exposure to air. An uncommon species found in deciduous woodland.

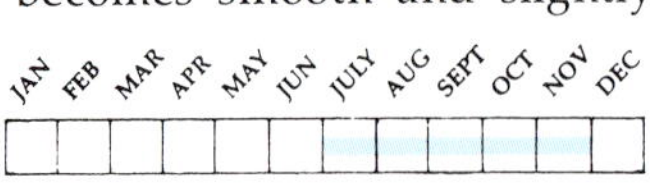

DOWNY BOLETE

Boletus subtomentosus

Cap diameter: 6-12cm
Height: up to 12cm
Characteristics: The cap is domed at first but later expands and flattens with age. The cap colour is grey-brown to yellowish buff. The surface is downy, as the English name suggests, and it may crack and split from the centre, exposing paler flesh below. The pores are yellow-brown and do not discolour when bruised. Spores olive-brown. The stem is slender and pale grey-brown and may be grooved towards the apex. Edible but not worth considering. The flesh is yellowish-brown, darker towards the cap.
Range and habitat: A common and widespread species in Britain and northern Europe. It grows among leaf-litter in deciduous woodlands.
Similar species: *B. porosprous* has a dull, yellowish-brown cap which cracks with age to reveal yellow flesh beneath. The pores are yellow and the stem is yellow-brown and ribbed. An uncommon species growing in deciduous woodlands, often under Oak.

RED-CRACKING BOLETE

Boletus chrysenteron

Cap diameter: 5-10cm
Height: up to 10cm
Characteristics: The cap is hemispherical at first but later expands and flattens with age. The cap surface is downy in young specimens but becomes smooth with age. The colour is clay to olive-brown and the surface readily cracks with age revealing the bright red flesh beneath. The pores are yellowish and may bruise blue. Spores olive-brown. The stem is yellowish staining and netted reddish towards the base. Edible but not to be recommended. The flesh is yellowish.
Range and habitat: An extremely common and widespread species. It grows in leaf litter in deciduous woodland.
Similar species: *B. calopus* has a flattened, clay-brown cap, yellow pores and a reddish stem covered in a network of white veins. The flesh is yellowish but turns bluish-white on exposure to air. An uncommon species which grows in deciduous woodland, especially under Beech and Oak.

BOLETUS PRUINATUS

Cap diameter: 4-8cm
Height: up to 10cm
Characteristics: The cap is domed at first but later expands and flattens with age. The colour is dark chestnut-brown but this fades to reddish around the margin with age. The cap surface has a powdery covering which is easily rubbed off. The pores are greenish-yellow and may bruise blue. Spores olive-brown. The stem is usually slightly swollen. It is yellow in colour, flushing red towards the base. Edible and considered good by some people. The flesh is yellowish-white, darker towards the stem base and turning bluish on exposure to air.
Range and habitat: A distinctly local and uncommon species throughout the region. It grows in leaf litter in deciduous woodland, especially under Oak and Beech.
Similar species: *B. lanatus* has a flattened buff-brown cap, yellow-orange pores and a yellow stem which reddens towards the apex. An uncommon species which grows in deciduous woodland, especially under Birch.

RED-CAPPED BOLETE

Boletus versicolor
Cap diameter: 3-6cm
Height: up to 8cm
Characteristics: The cap is domed at first but later expands and flattens with age. The cap surface has a slightly felty appearance and is generally bright red in colour. The pores are lemon-yellow in colour but bruise blue. Spores olive-brown. The stem is slender and tapers towards the apex. It is yellowish-orange becoming flushed red towards the base. Edible but not worth considering. The flesh is yellowish-buff in the cap staining orange down the stem.
Range and habitat: Widespread but a rather local species. It grows along grassy rides and clearing in deciduous woodland, especially under Oak.
Similar species: *B. appendiculatus* has a flattened and often irregular shaped cap. The cap colour is reddish-tan, the pores are yellow and the stem is yellowish. An uncommon species, less so in southern England, which grows in deciduous woodland, especially under Oak.

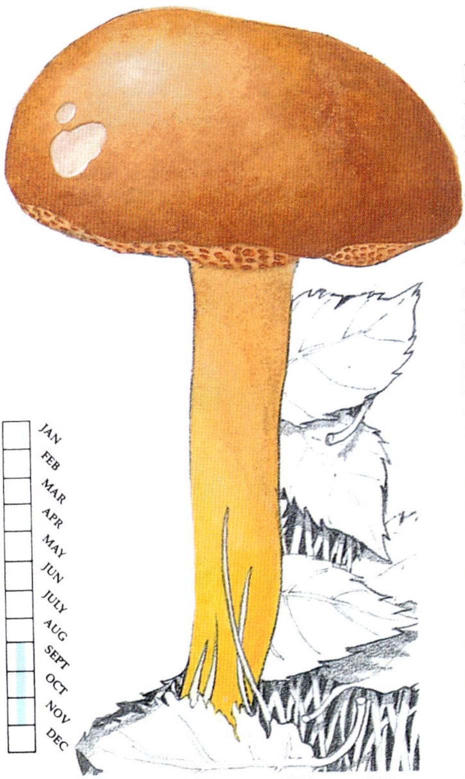

PEPPERY BOLETE

Boletus piperatus

Cap diameter: 4-10cm
Height: up to 10cm
Characteristics: At first the cap is domed but this expands and flattens somewhat irregularly with age. The cap surface is slightly sticky in damp weather and shiny when dry. The cap colour is orange-brown to tan. The pores are a rusty-brown colour. Spores olive-brown. The stem is slender and orange-brown in colour becoming yellow towards the base. Edible but best avoided. The flesh is yellowish becoming deeper yellow towards the base of the stem. As the English name suggest, it has a peppery taste.
Range and habitat: A common and widespread species throughout the region. It grows in deciduous woodlands, usually in association with Birch trees.
Similar species: *B. queletii* has a flattened, 5-15cm diameter tan cap. The pores are orange and the stem is yellowish-orange, flushing darker towards the base. An uncommon species found under Beech or Oak.

BROWN BIRCH BOLETE

Leccinum scabrum

Cap diameter: 5-15cm
Height: up to 20cm
Characteristics: The cap is domed at first but later expands and flattens with age. The cap is slightly sticky in damp weather but is otherwise dry. The cap colour is muddy-brown to buff-tan. The pores are creamy but bruise orange-brown. Spores rich brown. The slender stem tapers towards the apex. It is white, covered with darker scales. Edible but not worth considering. The flesh is white.
Range and habitat: A common and widespread species throughout the region. It grows most frequently in association with Birch trees.
Similar species: *L. variicolor* is similar but has a 5-10cm diameter brown cap blotched irregularly with grey. The pores are creamy-white and they bruise pink. The stem is whitish and covered in darker scales. The flesh is buffish-white but intense green near the stem base. A rather uncommon species, growing in association with Birch trees.

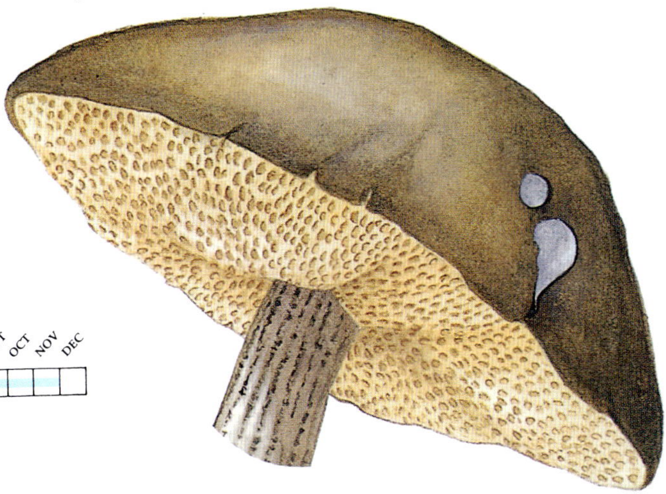

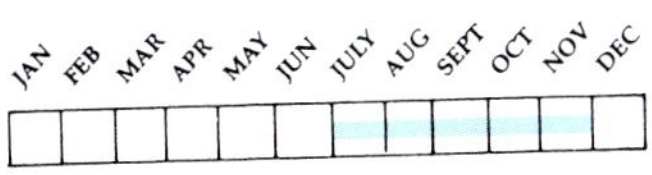

LECCINUM QUERCINUM

Cap diameter: 5-15cm
Height: up to 20cm
Characteristics: The cap is hemispherical at first. It expands with age but always remains extremely domed. The cap surface is slightly scaly but becomes smooth. It is orange-brown to tan in colour. The pores are creamy-yellow. Spores rich brown. The stem tapers towards the apex. It is whitish in colour, marked with a network or reddish, scaly veins. Edible and considered good by many people. Flesh whitish, staining grey or pink on exposure to air.
Range and habitat: A local and uncommon species throughout the region. It grows in association with Oak trees.
Similar species: The Yellow-cracking Bolete, *L. crocipodium*, has a brown cap which cracks to reveal yellowish flesh beneath. The pores are yellow and the swollen stem is buff covered in darker scales. The flesh is white but rapidly darkens on exposure to air. An uncommon species throughout the region, growing under Oaks.

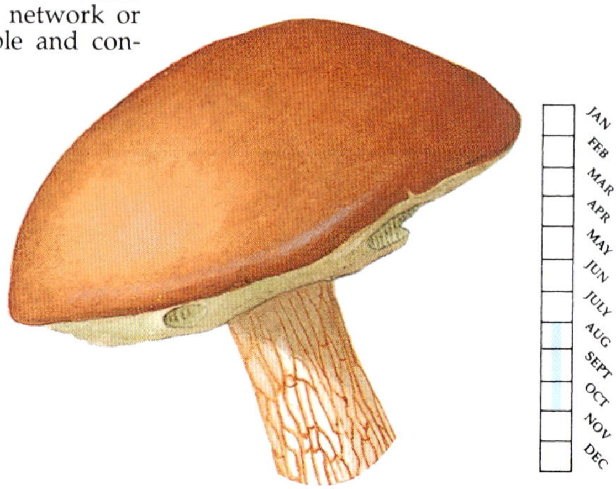

ORANGE BIRCH BOLETE

Leccinum versipelle
Cap diameter: 5-20cm
Height: up to 25cm
Characteristics: The cap is domed at first but later expands and flattens with age, still with a rather rounded appearance. The cap is rather downy at first but later becomes smooth. The colour is orange-tan, sometimes paler yellow around the margin. The pores are greyish in young specimens becoming yellowish with age. Spores rich brown. The stem is often swollen towards the base and is whitish, covered in a network of dark scales. Edible and good. Flesh white, staining dark green towards the stem base.
Range and habitat: A common and widespread species throughout the region. It grows almost exclusively in association with Birch.
Similar species: *L. aurantiacum* has a 5-15cm diameter, orange-brown cap, yellowish pores and a stout, whitish stem marked with darker scales. The flesh is whitish, staining darker. An uncommon species which grows with Birch and Aspen.

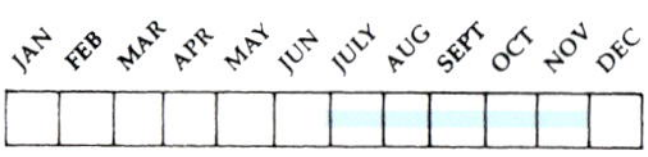

LARCH BOLETE

Suillus grevillei

Cap diameter: 5-10cm
Height: up to 10cm
Characteristics: At first the cap is a rounded conical shape but later it expands and becomes rather flattened. The cap surface is sticky in damp weather but otherwise dry and shiny. The colour is yellowish-orange to tan. The pores are yellow but bruise orange. Spores ochre. The stem is yellow above the whitish ring and orange, netted with darker scales, below. Edible but not worth considering. The flesh is yellow, darkening towards the stem base.
Range and habitat: A widespread species which is common in suitable habitats. It grows exclusively in association with Larch.
Similar species: *Aureoboletus cramesinus* has a sticky, reddish cap which is radially streaked with darker lines. The pores are yellow and the stem is yellow, staining reddish towards the base. An uncommon species which is found in deciduous woodland in southern England.

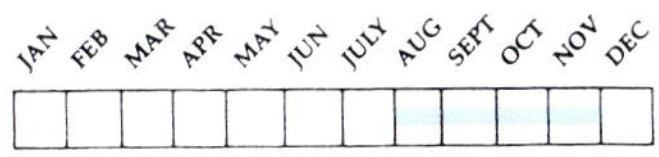

SUILLUS AERUGINASCENS

Cap diameter: 5-10cm
Height: up to 10cm
Characteristics: At first the cap is domed but it later expands irregularly and flattens. The surface is usually extremely sticky and the cap colour is olive-brown to buff. The pores are buffish-brown but bruise grey-green. Spores rich brown. The stem is rather slender and buffish-brown in colour. It bears a whitish ring. Edible but not worth considering. The flesh is pale-buff but may acquire a bluish flush on exposure to air.
Range and habitat: A rather uncommon and local species throughout the region. It grows almost exclusively under Larch trees.
Similar species: The Old Man of the Woods, *Strobilomyces floccopus*, has a 5-10cm diameter cap which is grey-brown in colour and covered in large, felty scales giving it an appearance rather like a pine-cone. The pores are grey and the long, slender stem is covered in grey-brown scales. An uncommon species, growing in both coniferous and deciduous woodland.

SLIPPERY JACK

Suillus luteus

Cap diameter: 2-10cm
Height: up to 12cm
Characteristics: The cap is domed at first but later expands and flattens rather irregularly. The cap surface is extremely sticky and slimy, hence the English name, but shiny when dry. The colour is orange-tan to muddy-brown. The pores are yellow-buff in colour. Spores ochre. The stem is pale brown, staining darker with age, and it is dotted with dark spots above the large, white ring. An edible species but not worth considering. The flesh is white, sometimes darkening towards the base of the stem.
Range and habitat: A common and widespread species throughout the region. It grows among fallen needles in coniferous woodland.
Similar species: *S. tridentinus* has an irregular, flattened cap, 5-10cm in diameter, which is orange-brown and covered in darker scales. The pores are orange and the stem is reddish brown. An uncommon species which grows under Larch.

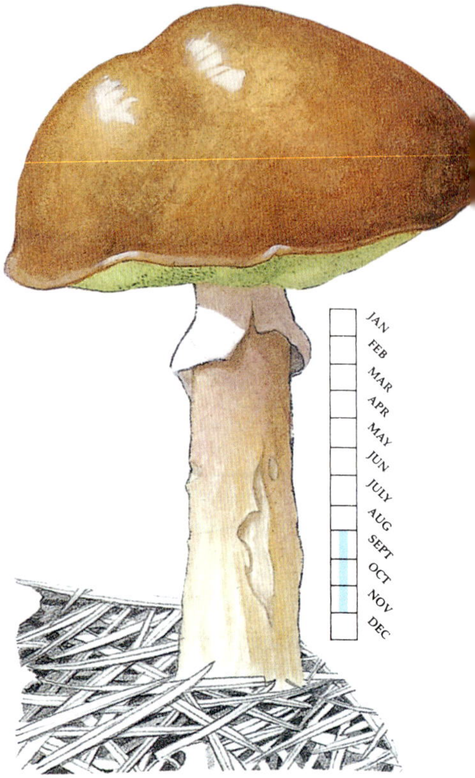

SUILLUS VARIEGATUS

Cap diameter: 6-12cm
Height: up to 12cm
Characteristics: The cap is almost hemispherical at first but later expands and flattens with age. The surface is slightly scaly and sticky in damp weather. The cap colour is olive-brown but covered with darker brown scales. The pores are olive-brown becoming darker brown with age. Spores rich brown. The stout and swollen stem is yellowish in colour, becoming darker towards the base. Edible and considered good by some people. The flesh is yellowish-white, sometimes tinged with blue.
Range and habitat: A fairly common and widespread species throughout the region. It grows in coniferous woodland on well-drained soils.
Similar species: *Gyroporus castaneus* is an uncommon species which grows in Oak woodland. The cap is chestnut-brown and 5-10cm in diameter. The pores are yellow and the stem is brown and slightly swollen. An edible and delicious species which has rather brittle flesh.

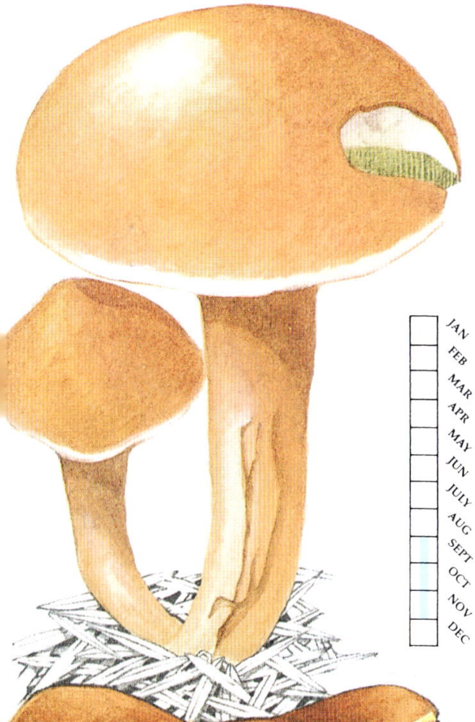

SUILLUS BOVINUS

Cap diameter: 5-10cm
Height: up to 10cm
Characteristics: The cap is a rounded conical shape at first but later expands and becomes rather domed. The cap surface is rather sticky. The colour is orange-brown with a distinct, white margin. The pores are orange-brown to olive-tan in colour. Spores olive-brown. The stem is often rather tapering and orange-brown in colour. Edible and considered good by many people. The flesh is yellowish-white becoming darker towards the stem base.
Range and habitat: A common and widespread species throughout the region in suitable habitats. It grows among pine needles in coniferous woodland, especially in association with Scot's Pine.
Similar species: *S. flavidus* has a flattened, slightly umbonate cap which is yellow-brown in colour and 2-5cm in diameter. The pores are yellow and the stem is whitish with a brown ring. An uncommon species found in Scottish coniferous woodland.

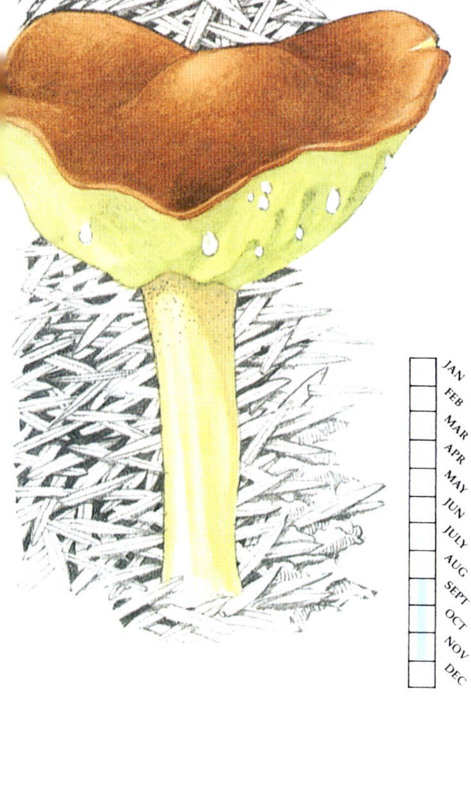

SUILLUS GRANULATUS

Cap diameter: 4-8cm
Height: up to 10cm
Characteristics: The cap is hemispherical at first but later expands and flattens to an irregular shape. The surface is smooth and sticky and the cap colour is deep-red to orange-brown. The pores are yellow and exude milky drops in damp weather. Spores ochre. The slender stem is yellowish and the apex is covered in granules which also exude milky droplets. Edible and considered good by some people. The flesh is whitish-yellow, staining more intense yellow towards the base of the stem.
Range and habitat: A common and widespread species throughout the region. It grows among pine needles in coniferous woodland.
Similar species: *S. collinitis* (=*S. fluryi*) has a 5-10cm diameter, brown cap, covered with darker streaks. The pores are yellow and the stem is buff with darker flecks. It grows in coniferous woodland but is not found in Britain. Edible but not to be recommended.

Paxillus involutus

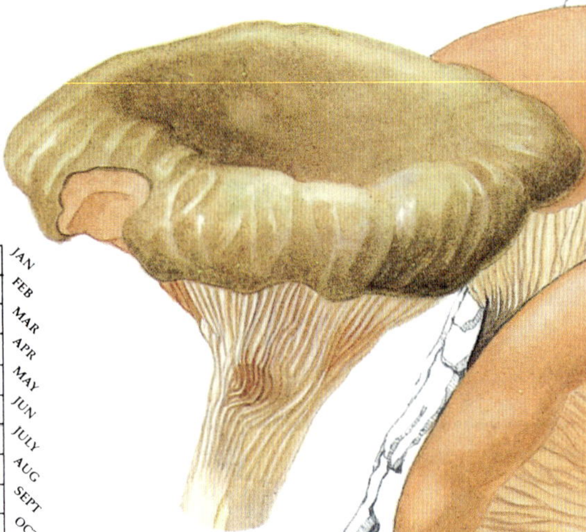

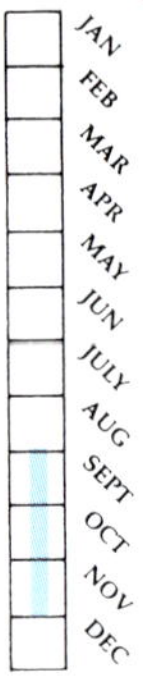

Paxillus atromentosus

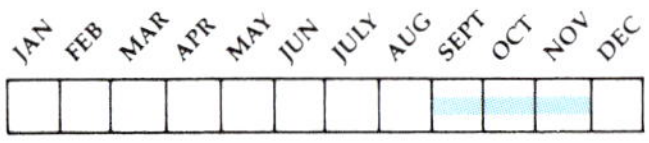

BITTER BOLETE
Tylopilus felleus
Cap diameter: 6-12cm
Height: up to 12cm
Characteristics: The cap is domed at first but later expands and flattens with age, often with a rather wavy, irregular margin. The surface is rather downy at first but becomes smooth with age. The cap colour is chestnut to reddish-brown. The pores are clay-brown but bruise darker brown. Spores pinkish. The stout stem is often swollen towards the base. It is buff-brown in colour and covered in a network of dark-brown veins. Not edible. The flesh is creamy-white with an extremely bitter taste as the English name suggests.
Range and habitat: A widespread and locally common species throughout the region. It grows in both coniferous and deciduous woodland.
Similar species: *Porphyrellus porphyrosporus* (=*pseudoscaber*) has dull brown, 5-15cm diameter cap. The pores and swollen stem are brown. An uncommon species which grows in mixed woodlands.

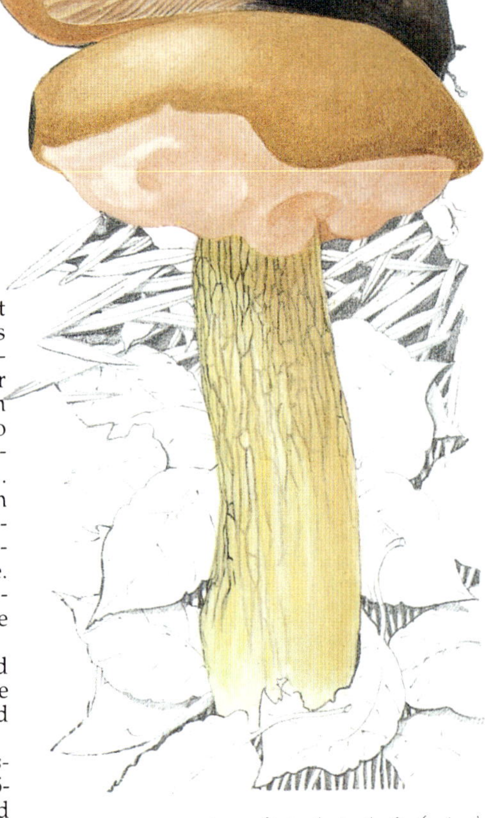

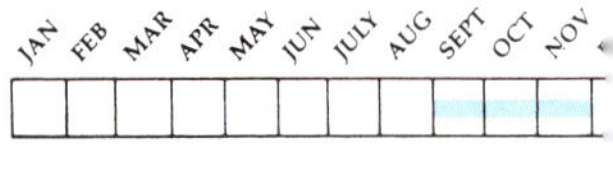

BROWN ROLL-RIM

Paxillus involutus

Cap diameter: 5-15cm

Height: up to 10cm

Characteristics: At first the cap is a flattened dome with a depressed centre and inrolled margin. It expands with age, the centre becoming more depressed and the fungus appearing funnel-shaped. The margin remains inrolled, as the English name suggests. The cap margin is downy and the surface is sticky in wet weather. The cap colour is orange-tan. The gills are decurrent and orange-buff in colour, bruising a chestnut colour. Spores ochre. The stem is the same colour as the cap and stains and bruises chestnut. Poisonous and to be avoided. Older books may indicate, incorrectly, that this species is edible after boiling. Flesh yellowish.

Range and habitat: An extremely common and widespread species throughout the region. It grows mainly in deciduous woodland and especially in association with Oak and Birch.

Similar species: *P. atrotomentosus* is much stouter. Grows under conifers.

PAXILLUS ATROTOMENTOSUS

Cap diameter:10-30cm

Height: up to 15cm

Characteristics: At first the cap is domed. Later it expands and flattens to form an irregular and uneven cap. The cap centre is depressed and the margin is distinctly inrolled. The cap colour is orange-tan to grey-brown, usually blotched irregularly with paler buff marks. The gills are creamy-white to yellow-buff and are strongly decurrent. Spores brown. The stem is extremely stout and robust and is covered in a dense coat of dark-brown to black, velvety hairs. Not edible. The flesh is creamy-white.

Range and habitat: A widespread species which is occasionally locally common in suitable habitats. It grows on rotting stumps and trunks in coniferous woodland and is often found in clumps.

Similar species: *P. panuoides* has a fan-shaped, buff cap, 5cm across, which is attached to conifer wood by a tiny, flattened stem. The gills are decurrent and buffish-yellow. A rather uncommon species. Not edible.

CHROOGOMPHUS RUTILUS

Cap diameter: 5-15cm

Height: up to 15cm

Characteristics: At first the cap is hemispherical but it later expands and flattens with a distinct umbo and the margin slightly inrolled. The surface is sticky in damp weather but dries shiny. The cap colour is reddish-brown to tan. The gills are widely spaced, decurrent and buffish-brown at first but darkening with age. Spores dark brown. The stem is yellowish in colour, staining reddish at the apex and bright yellow towards the base. There is the remains of a slight ring near the apex. Edible but best avoided. Flesh creamy-buff, staining yellow towards the base of the stem.

Range and habitat: A common and widespread species throughout the region. It grows in coniferous woodland.

Similar species: *Gomphidius glutinosus* has a sticky, brown cap, 5-10cm in diameter, decurrent white gills and a slender stem which is white, staining yellow towards the base. An uncommon species which grows under conifers.

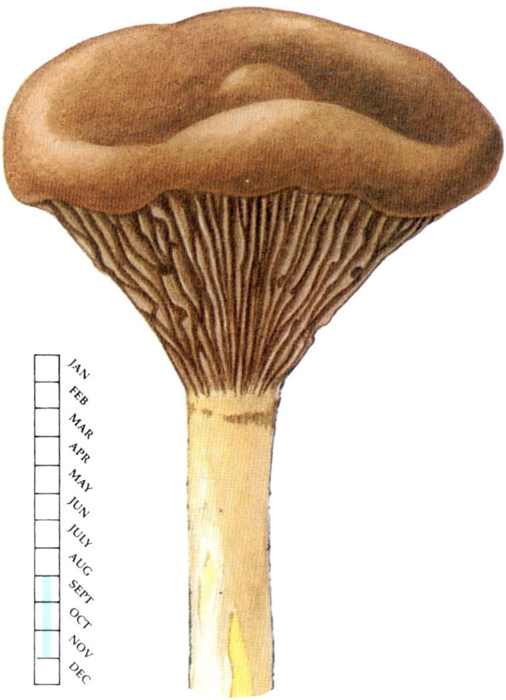

Stereum hirsutum

BRACKET FUNGI

In terms of their appearance, size, shape and habitat, the bracket fungi are among the most diverse of all fungus groups.

Of those with a conventional bracket shape, some are large, such as the Dryad's Saddle, *Polyporus squamosus*, while others, like *Bjerkandera adusta*, are much smaller but grow in huge, overlapping masses.

The group also includes such unusual species as the Earpick Fungus, *Auriscalpium vulgare*, and the Wood Hedgehog, *Hydnum repandum*, both of which have spines underneath the cap.

Many species of bracket-related fungi are of significant economic importance. The Horn of Plenty, *Craterellus cornucopioides*, is eaten in large quantities but it is the Chanterelle, *Cantharellus cibarius*, that is most prized. Several other species of bracket fungi cause rot in the heart-wood of their host trees. Some of these can cause devastating losses of timber to the forestry industry as well as a deterioration in timber quality.

EAR-PICK FUNGUS

Auriscalpium vulgare

Cap diameter: 1-2cm
Height: up to 9cm
Characteristics: An intriguing fungus which resembles a Victorian earpick. The cap is kidney-shaped or semi-circular. The surface is leathery and slightly hairy. The cap colour is dark-brown to black. The underside of the cap is covered in spines. These are dark-brown at first but later acquire a white dusting of spores with maturity. The stem is attached laterally to the cap and becomes swollen towards the base. It is dark-brown in colour and becomes increasingly hairy towards the base. Not edible.
Range and habitat: A common and widespread species throughout the region in suitable habitats. It grows on buried pine cones and, because of its colouring, is often extremely difficult to spot.
Similar species: Because of the habitat and appearance of this species, it is unlikely to be confused with any other.

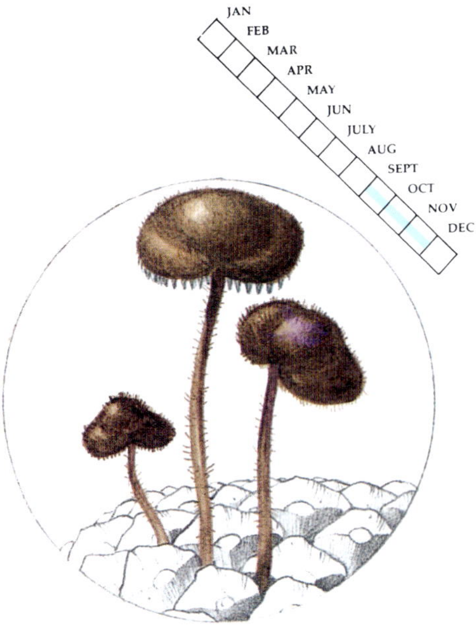

CANTHARELLUS INFUNDIBULIFORMIS

Cap diameter: 2-6cm
Height: up to 8cm
Characteristics: At first the cap is flattened with a depressed centre. Later it expands and becomes funnel-shaped and often mis-shapen with age with an irregular and wavy margin. The cap colour is dark, chocolate-brown. The gill-like wrinkles are yellowish-brown and have distinct cross-veins. Spores cream. The stem is stout and robust, tapering and flattened towards the base. It may have a grooved surface. The stem colour is bright orange-yellow. Edible and considered good by many people. The flesh is yellowish and slightly bitter.
Range and habitat: A widespread species which is fairly common throughout Britain and northern Europe. It grows in deciduous woodland and, to a lesser extent, under conifers.
Similar species: *C. lutescens* (=*C. infundibuliformis* var. *lutescens*) has a dull orange-yellow fruiting body as well as stem and a more northern distribution.

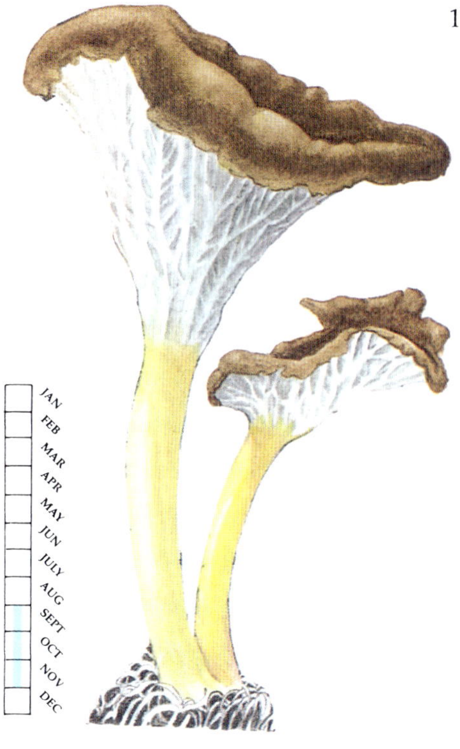

CHANTERELLE

Cantharellus cibarius

Cap diameter: 4-10cm
Height: up to 10cm
Characteristics: At first the cap is a flattened-dome with a wavy margin. Later it expands and becomes funnel or trumpet-shaped with an irregular and wavy margin. The cap colour is a deep egg-yolk yellow to orange and the surface is smooth. The gill-like wrinkles and folds are decurrent, irregular and yellow in colour. Spores cream. The stem tapers towards the base and is the same colour as the cap. An edible and delicious species. This is one of the most highly-prized of all the edible fungi. The flesh is whitish and, when fresh, has a distinct smell of apricots.
Range and habitat: A widespread species throughout the region which is locally common in some areas. It grows mainly in deciduous woodland but sometimes also under conifers.
Similar species: This species' shape, colour and apricot smell make it extremely distinctive and difficult to confuse with any other.

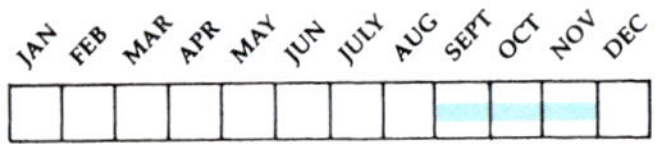

HORN OF PLENTY

Craterellus cornucopioides

Cap diameter: 4-8cm
Height: up to 10cm
Characteristics: The cap is funnel-shaped but is extremely irregular in appearance with a highly folded and wavy margin. The surface is slightly downy and is thin and leathery to touch. When wet the surface is dark brown but it dries paler grey-brown. The spore-producing underside of the cap lacks gills or gill-like wrinkles, although it may be slightly ribbed. It is grey and is continuous with the stem which is short and stout, tapering towards the base. Spores cream. Edible and good, being especially suited to drying. The white flesh has a pleasant taste and smell.
Range and habitat: A widespread species which may be locally abundant. It grows in deciduous woodlands, especially under Beech.
Similar species: *Cantharellus cinereus* is similar in appearance but has whitish, gill-like wrinkles on the underside of the cap. A rather uncommon species, found under Beech.

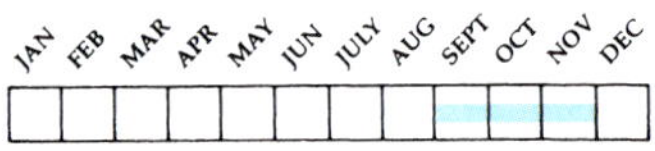

WOOD HEDGEHOG

Hydnum repandum

Cap diameter: 5-15cm
Height: up to 8cm
Characteristics: At first the cap is flattened with a depressed centre and inrolled margin. It later expands and flattens to an irregular shape but retains the slightly depressed centre. The surface is velvety at first but smooth later and the cap colour is creamy-buff. The underside of the cap is covered in densely-packed spines which are pinkish-buff in colour. Spores white. The stem is short and stout and is often distorted and off-centre. The stem colour is whitish-buff but stains darker towards the base. Edible and considered good by many people. The flesh is white.
Range and habitat: A widespread species which may be locally common. It grows in both coniferous and deciduous woodland.
Similar species: Species of *Sarcodon* also have a spiny underside to the cap. They are generally darker brown with scaly caps and grow in coniferous woodland.

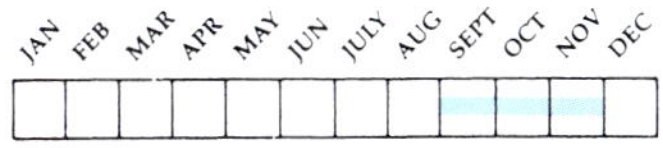

EARTH-FAN

Thelephora terrestris

Diameter: 3-6cm
Characteristics: As the English name suggests, this fungus is fan-shaped with a feathery and frayed margin giving mature specimens an extremely shaggy appearance. The surface is covered in radiating fibres and is densely hairy. The colour of the upper surface is clay or buff, darkening to dark brown with age. The under surface is smooth or sometimes slightly wrinkled and pale brown in colour. Spores purple-brown. Not edible.
Range and habitat: A common and widespread species in Britain and northern Europe. It grows among mosses and fallen pine needles in coniferous woodland and on heathlands. It often forms densely-packed and overlapping masses which are initially difficult to spot on the woodland floor.
Similar species: *T. palmata* has the fan much divided into palmate fronds, joined and fused at the base. The fungus smells strongly of garlic. It grows in coniferous woodland and is rather local and uncommon.

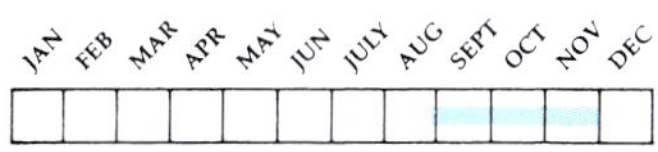

HAIRY STEREUM
Stereum hirsutum
Bracket: 3-8cm across
Characteristics: The fungus forms irregular semi-circular or elongated brackets which are tough and leathery. The upper surface has a fine covering of hairs and is generally orange-brown in colour, often zoned with grey or darker brown and with a paler margin. The smooth under surface is yellow but fades to pale buff with age. The fungus does not discolour when bruised. Spores white. Not edible.
Range and habitat: An extremely common and widespread species throughout the region. It grows in deciduous woodland often forming large, overlapping clumps on stumps, logs and fallen branches and twigs.
Similar species: *S. rugosum* has buffish-brown, slightly encrusting brackets which bleed red when cut in fresh specimens. The brackets dry greyish with age. It is common and widespread throughout the region, growing on stumps, logs and fallen wood of deciduous trees, especially Hazel.

JAN	FEB	MAR	APR	MAY	JUN	JULY	AUG	SEPT	OCT	NOV	DEC

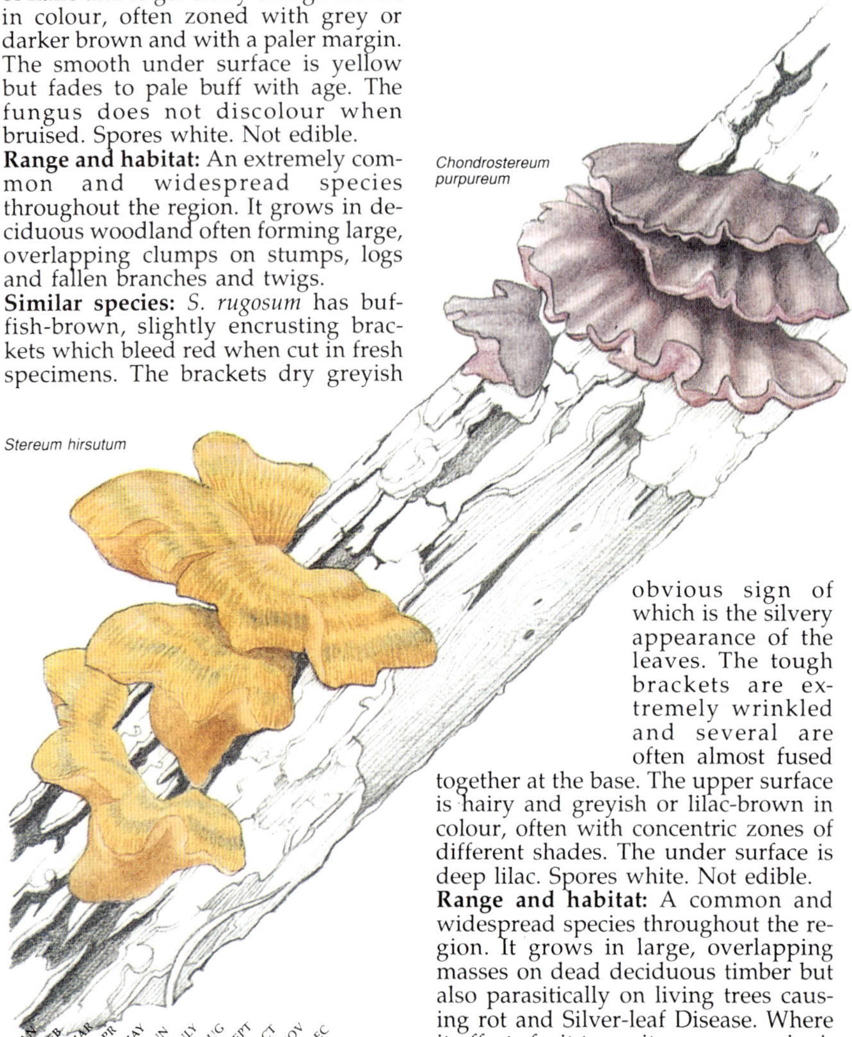
Chondrostereum purpureum
Stereum hirsutum

JAN	FEB	MAR	APR	MAY	JUN	JULY	AUG	SEPT	OCT	NOV	DEC

SILVER-LEAF FUNGUS
Chondrostereum purpureum
Bracket: 3-8cm across
Characteristics: This fungus causes Silver-leaf Disease in fruit trees, the most obvious sign of which is the silvery appearance of the leaves. The tough brackets are extremely wrinkled and several are often almost fused together at the base. The upper surface is hairy and greyish or lilac-brown in colour, often with concentric zones of different shades. The under surface is deep lilac. Spores white. Not edible.
Range and habitat: A common and widespread species throughout the region. It grows in large, overlapping masses on dead deciduous timber but also parasitically on living trees causing rot and Silver-leaf Disease. Where it affects fruit trees, it may cause significant damage to orchards.
Similar species: *Trichaptum abietinum* has leathery, overlapping brackets which are 1-3cm across. The upper surface is distinctly zoned with purple, grey and brown and is slightly hairy. Common and widespread on stumps and trunks of conifers.

SHAGGY POLYPORE

Inonotus hispidus

Bracket: 10-20cm across

Characteristics: A distinctive species which forms large brackets up to 10cm thick. The upper surface is covered in a dense layer of shaggy hairs and the colour is orange-tan to reddish-brown, becoming much darker with age. The pores on the under surface are yellowish-orange at first but darken with age. Spores yellow. Not edible.

Range and habitat: A common and widespread species in Britain and northern Europe. It usually grows singly on the trunks of deciduous trees, especially on Ash and Elm.

Similar species: *I. dryadeus* forms large, semi-circular brackets up to 60cm across and 10cm thick. The greyish upper surface darkens with age and the pores are whitish, reddening with age. Rather uncommon, growing on the trunks of Oaks. *I. radiatus* has brackets 5-10cm across which are brownish, darkening with age. Common and widespread, found on deciduous trees such as Hazel and Alder.

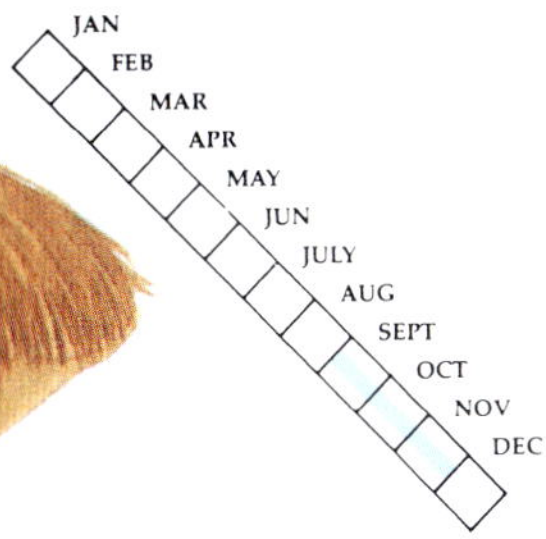

LACQUERED BRACKET

Ganoderma lucidum

Bracket: 10-20cm

Characteristics: The hard bracket is semi-circular or kidney-shaped, often with distinct lobes. The surface is very smooth and shiny as if varnished or lacquered. The upper surface is orange-brown or reddish, darkening with age, and has concentric grooves and zones. The margin is often paler yellow. The pores on the under surface are creamy-yellow in colour. Spores brown. Stem short and stout. Attached laterally to the bracket. It is the same colour as the cap. Not edible.

Range and habitat: A widespread species throughout the region which is generally not common. It grows on trunks and logs of deciduous trees, usually near the base.

Similar species: *G. resinaceum* (also called Lacquered Bracket) has a semi-circular bracket, 10-40cm across, which is reddish brown with concentric zones and grooves. The pores on the under surface are pale-buff. An uncommon species which grows on stumps of Oak.

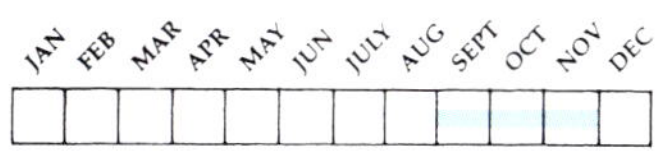

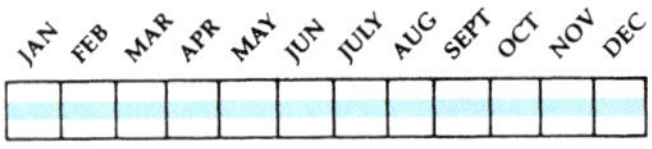

ARTIST'S FUNGUS

Ganoderma europeaeum (=*G. applanatum*)

Bracket: 10-50cm across

Characteristics: The bracket is usually semi-circular in shape and 3-6cm thick. Although it may resemble an artist's palate, the English name derives from the fact that the under surface may be drawn on. The upper surface is knobbly, hard and smooth and comprises layers of concentric growth rings. The colour of the upper surface is reddish-brown with a whitish margin. The pores on the under surface are pale brown but bruise reddish-brown. The spores stain the tree below reddish-brown. Not edible. The flesh is reddish-brown and tastes bitter.

Range and habitat: A widespread species in the region but rather uncommon in Britain. It grows on the trunks and branches of deciduous trees, especially on Beech.

Similar species: *G. adspersum* is similar in appearance but has a rich brown upper surface and buff-brown under surface. It grows on the trunks on deciduous trees.

HOOF FUNGUS; TINDER FUNGUS

Fomes fomentarius

Bracket: 5-40cm across

Characteristics: An aptly-named species which is shaped like a horse's hoof. The bracket is up to 25cm thick and is hard and woody. The upper surface comprises concentric zones and grooves and is covered in a greyish-brown horny layer. The margin may be paler. The pores are pale brown. Spores white. Not edible. The flesh is brown and extremely tough.

Range and habitat: A widespread species in the region which is rare and local in southern England but more frequent further north and in Scotland in particular. It generally grows on Birch trunks in Scotland but on Beech in England. Several may occur on the same trunk, causing the timber to rot.

Similar species: The Root Fomes, *Heterobasidion annosum*, is common and causes widespread damage to conifer plantations. The upper surface of the thick bracket is reddish-brown and knobbly and the under surface is white. Grows near the tree base.

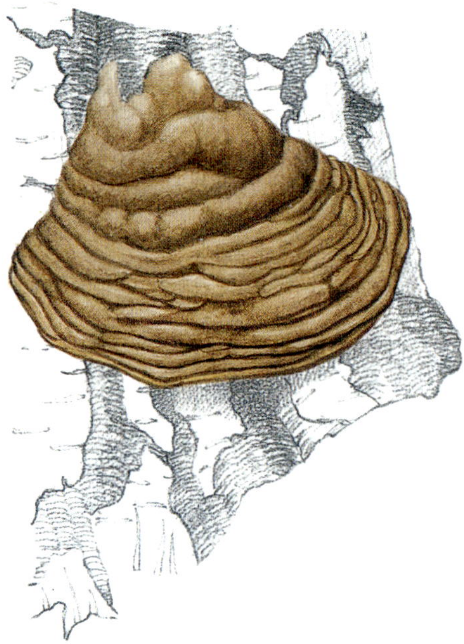

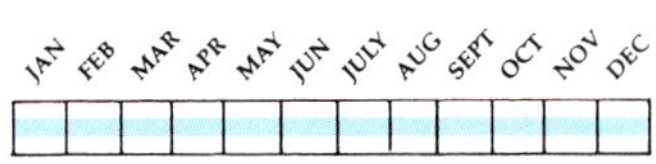

BIRCH POLYPORE

Piptoporus betulinus

Bracket: 10-20cm across

Characteristics: Young specimens have a smooth, rounded appearance but with age the bracket expands and flattens. Mature specimens are rather flattened and leathery, from 3-6cm thick, and have a beautifully rounded margin. The upper surface is buffish-white, and later acquires concentric zones of pinkish-orange. The margin is usually paler than the rest of the upper bracket and the surface may crack in older specimens. The pores on the underside are whitish and do not discolour. Spores white. The whole bracket is extremely long-lasting. Not edible. The flesh is white and spongy, formerly serving many useful purposes.

Range and habitat: A common and widespread species throughout the region. As the English name suggests, this species grows exclusively on the trunks of Birch trees.

Similar species: The habitat and characteristic appearance make it difficult to confuse with any other species.

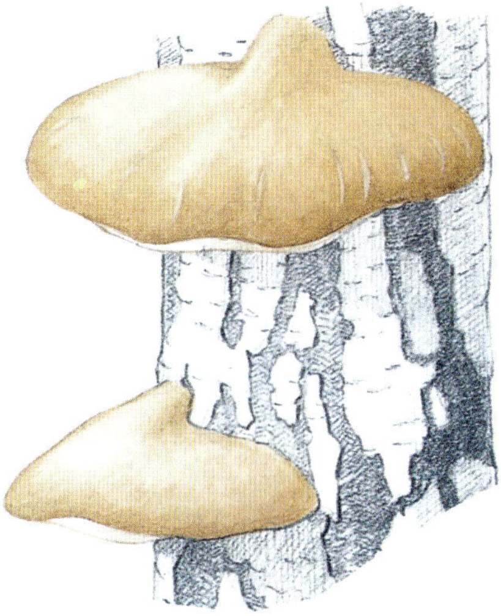

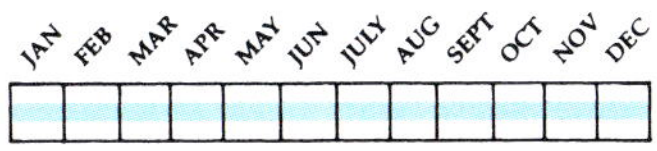

SULPHUR POLYPORE; CHICKEN OF THE WOODS

Laetoporus sulphureus

Bracket: 10-45cm across

Characteristics: An elegant and unmistakable species. Each bracket is an irregularly semi-circular or tongue- or fan-shaped. The margin is often lobed and the whole bracket is knobbly, distorted and up to 10cm thick. It often grows in large, tiered and overlapping masses which may appear fused. The upper surface is soft and felty and bright, sulphur yellow in young specimens. It fades to creamy-white with age. The pores on the under surface are similarly coloured. Spores white. Edible and considered good by many people. The flesh is soft and whitish but becomes crumbly with age.

Range and habitat: A widespread species which is locally common in some areas. It grows on deciduous trees, especially Sweet Chestnut and Oak, causing heart rot of the tree.

Similar species: The size, appearance and bright yellow colour of this species make it difficult to confuse with any other.

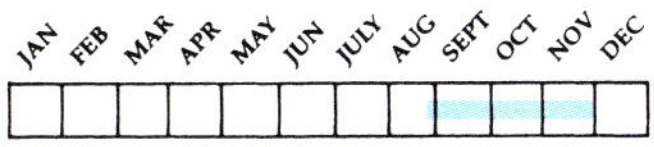

DRYAD'S SADDLE

Polyporus squamosus

Bracket: 10-50cm across

Characteristics: A large and impressive species. At first the bracket is almost circular but later it expands and flattens becoming semi-circular or fan-shaped. The upper surface is orange-brown to buff and is covered with dark brown, fibrous scales arranged in vaguely concentric rings. The pores on the under surface are creamy-white. Spores white. The stem is short and stout, darkening towards the base. Edible but not to be recommended. Flesh white and soft in young specimens with a mealy smell.

Range and habitat: A common and widespread species throughout the region. It grows on the stumps and trunks of deciduous trees. Both living and dead trees are affected.

Similar species: The appearance and large size of this species make it almost unmistakable. *P. floccipes* is similarly coloured but has a cap 2-8cm across and a distinct stem. It grows on fallen twigs of deciduous trees such as Oak and Hazel.

Polyporus floccipes

PSEUDOTRAMETES (=TRAMETES) GIBBOSA

Bracket: 5-15cm across

Characteristics: The bracket is semi-circular or kidney-shaped. The upper surface is rather uneven and downy or hairy in young specimens. The colour of the upper surface is whitish, often with a thick, creamy margin. It may be flushed with pink or yellowish and often becomes stained green due to the presence of a film of algae growing on the surface. The pores on the underside are white and rather elongated. Spores white. Not edible. The flesh is corky and white.

Range and habitat: A common and widespread species in Britain and northern Europe. It grows on decaying stumps and logs of deciduous trees, especially on Beech.

Similar species: *Phellinus ignarius* has a thick, kidney-shaped cap which is extremely hard and woody. It is blackish-green on the upper surface with buffish pores on the underside. Generally rather uncommon and parasitic on deciduous trees, especially on Willows. Causes a white rot in the timber.

WINTER POLYPORE

Polyporus brumalis

Cap diameter: 4-8cm

Height: up to 5cm

Characteristics: Unlike most bracket fungi, this species has a distinct cap and stem. The smooth cap is round and greyish-brown in colour with a slightly inrolled margin. The pores are white and decurrent. Spores white. The stem is slender and tapering and is whitish in colour. Considered edible by some but not to be recommended. The flesh is white and tough.

Range and habitat: A widespread species in the region but generally rather local. It grows on fallen wood of deciduous trees.

Similar species: *P. ciliatus* is similar but has the stem marked with dark bands. *P. varius* has a kidney- or fan-shaped cap, up to 10cm across, sometimes with a short stem. The upper surface of the cap is buffish-yellow and the pores on the underside are white. It grows on dead and dying deciduous wood and especially on Willow.

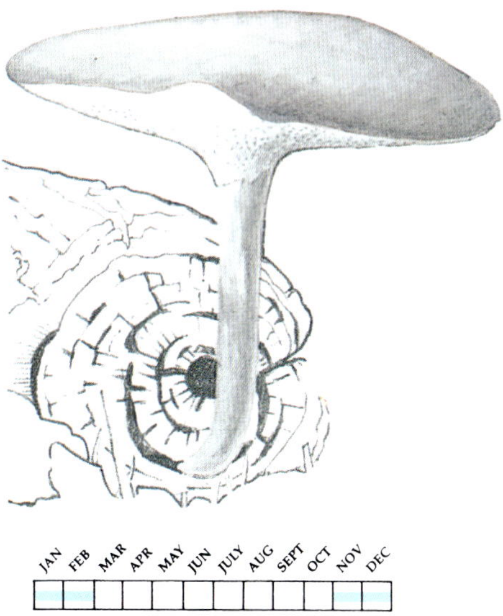

LENZITES BETULINA

Bracket: 4-8cm across.

Characteristics: The bracket is semicircular or fan-shaped and up to 2cm thick. The upper surface is tough and leathery and often rather uneven. It is concentrically grooved and zoned with various colours including reddish-brown, buff, grey and white and the surface is slightly hairy and may be stained green with algae. The bracket margin is generally pale. The pores on the underside are extremely thin and elongated and consequently resemble gills. They are whitish and contrast with the reddish-brown flesh beneath them. Spores white. Not edible. The flesh is white and soft.

Range and habitat: A common and widespread species throughout the region. It grows on the stumps and trunks of deciduous trees and especially on Birch.

Similar species: The colour and unusual shape of the pores make this species difficult to confuse with other bracket fungi, except perhaps the Maze-gill, *Daedelea quercina,* which grows on Oak.

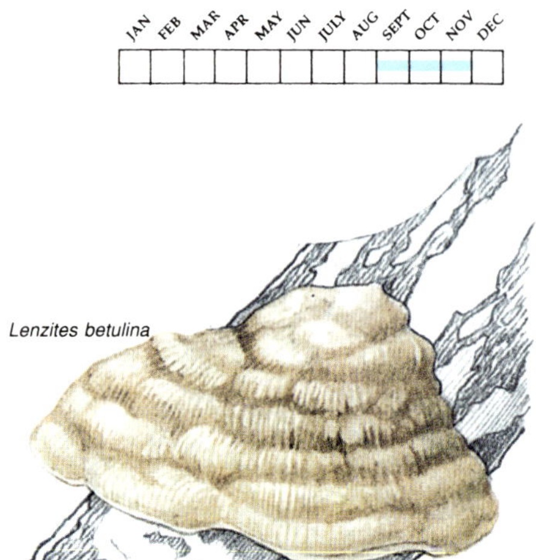

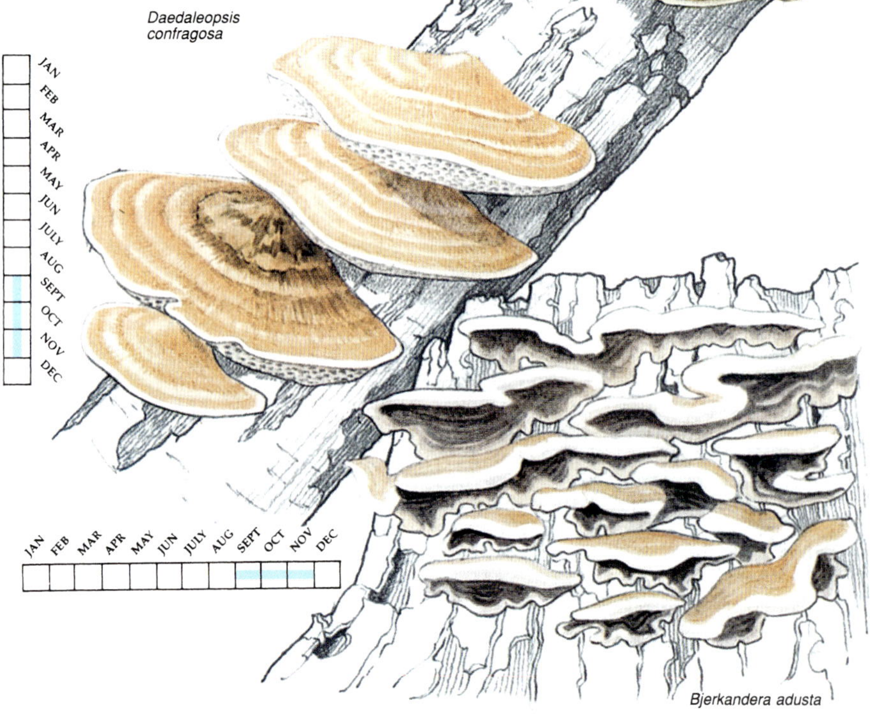

BLUSHING BRACKET

Daedaleopsis confragosa

Bracket: 4-8cm across

Characteristics: The bracket is often a beautifully rounded, semi-circular shape which may be up to 2cm thick. The upper surface is generally ridged and marked with concentric zones of colour, alternating between reddish-brown and dark brown. Some specimens may be buffish-tan. The margin is often noticeably paler than the rest of the upper surface. The pores on the underside are whitish but immediately stain red when bruised. Spores white. Not edible. The flesh is pinkish and has a corky texture.

Range and habitat: An extremely common and widespread species in Britain and northern Europe. It grows on living and dead wood of deciduous trees and especially on Willow. The brackets often grow in tiers on vertical branches.

Similar species: The red-bruising pores are a useful guide to this species' identity. Some specimens may have rather elongated pores resembling gills.

BURNT POLYPORE

Bjerkandera adusta

Bracket: 4-8cm across

Characteristics: The individual brackets are irregularly semi-circular but often grow in overlapping and fused groups and tiers. The bracket is around 0.5cm thick and markedly undulating. The upper surface is greyish-buff and may have a white margin. This darkens with age. The whole fungus tends to darken with age. The pores on the underside are greyish in young specimens but soon darken. Spores white. Not edible.

Range and habitat: A widespread and often abundant species throughout Britain and northern Europe. It grows on rotting stumps and branches of deciduous trees. It often forms large and extensive groups which sometimes cover the entire surface of the timber.

Similar species: A rather distinctive species but faded specimens may resemble the Silver-leaf Fungus, *Chondrostereum purpureum,* while bright brackets can be similar to the Many-zoned Polypore, *Coriolus versicolor.*

GIANT POLYPORE

Meripilus giganteus

Bracket: 10-40cm across

Characteristics: Individual brackets are irregularly fan-shaped but they grow in huge, overlapping and sometimes fused clumps resembling large rosettes. The upper surface of the bracket is fleshy and leathery and covered in radial streaks. The colour is buffish-brown overlaid with darker brown scales, often arranged in zones of alternating light and dark colour. The margin is usually extremely irregular and often torn or frayed. The pores on the underside are yellowish but bruise blackish. Spores white. There is sometimes a tiny stem. Not edible. The flesh is fibrous and white.

Range and habitat: A common and widespread species throughout the region which is difficult to overlook because of its size. It grows at the base of deciduous trees.

Similar species: *G. frondosa* is smaller with more elongated individual brackets. An uncommon species which grows at the base of Oaks and Beeches.

MAZE-GILL

Daedalea quercina

Bracket: 5-15cm across

Characteristics: The bracket is semi-circular to fan-shaped in outline and up to 5cm thick. It is tough and corky with the upper surface a buffish-green or grey-brown in colour. The surface is marked with concentric ridges and zones of different colours. The English name is extremely apt and derives from the whitish pores on the under-side which are greatly elongated and gill-like and arranged in the fashion of a maze. Spores white. Not edible. The flesh is buffish and corky.

Range and habitat: A common and widespread species throughout Britain and northern Europe. It grows on rotting stumps and trunks of deciduous trees, almost exclusively on Oak. Generally grows singly but may occur in sizeable, overlapping and tiered groups.

Similar species: The patterning of the pores on the underside are a good indicator to this species' identity. Some specimens may, however, superficially resemble *Lenzites betulina*.

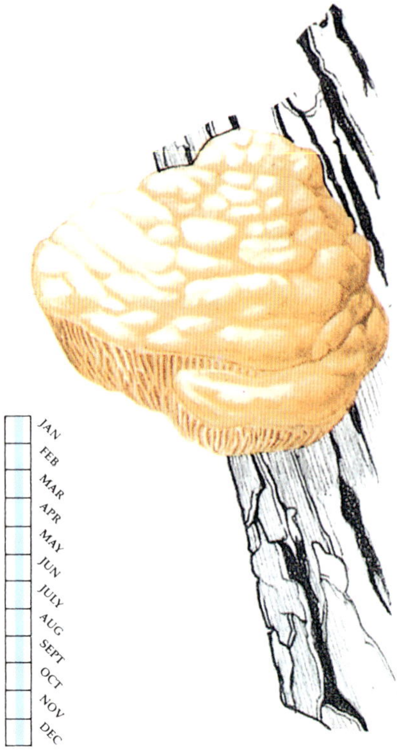

MANY-ZONED POLYPORE

Coriolus versicolor

Bracket: 5-10cm across

Characteristics: A frequently encountered species of bracket fungus. The individual bracket is roughly semi-circular in outline although the margin is often lobed. Generally grows in dense and overlapping tiered groups. The bracket is tough and leathery and the upper surface has a velvety texture. It is marked with concentric ridges and clearly-defined zones of alternating colours including, reddish-brown, grey, white and greenish-blue. The margin is invariably creamy-white. Pores on the under surface are white. Spores white. Not edible. Flesh tough and white.

Range and habitat: An extremely common and widespread species throughout the region. It grows in large groups on deciduous timber.

Similar species: This is an extremely variable species. However, the white margin is usually a good feature at all stages. See also *Bjerkandera adusta* which may have a pale margin in young specimens.

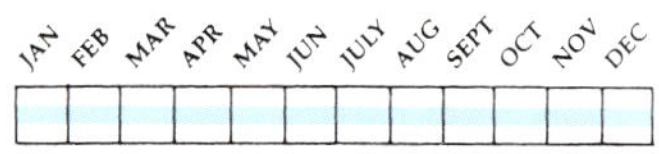

BEEFSTEAK FUNGUS
Fistulina hepatica
Bracket: 10-30cm across
Characteristics: An almost unmistakable species whose bracket is kidney- or tongue-shaped in outline and up to 7cm thick. The soft, fleshy surface is sticky and moist. The colour is blood-red, darkening with age. The combination of shape, colour and texture make the English name extremely appropriate. The pores are whitish but bruise and age dark red. Spores reddish-brown. Edible and considered good by some people. Flesh red. Oozes blood-like fluid when cut, adding to its flesh-like appearance.
Range and habitat: A widespread and locally common species throughout the region. It grows parasitically on Oak and occasionally on other species of deciduous trees, usually low down near the base of the trunk. The fungus causes a brown heart-rot in the timber, and the resulting stained wood is prized.
Similar species: The appearance, texture and colour of this species make confusion unlikely.

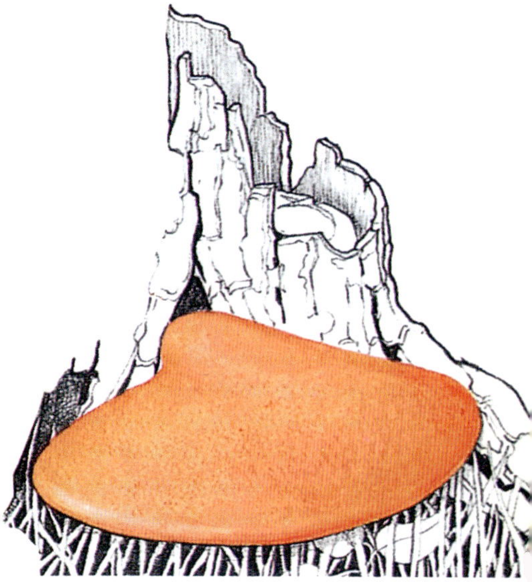

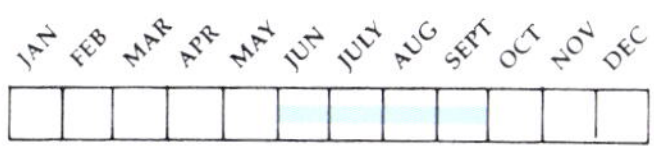

GIANT FAIRY CLUB

Clavariadelphus pistillaris

Height: 10-30cm

Characteristics: This extraordinary looking species resembles a large pestle. Some specimens are more swollen than others and may be up to 5cm thick at the widest point. The colour is yellow-orange to buff-brown but may blotch and bruise darker with age. The surface is generally smooth but may be wrinkled below the most swollen portion of the club. Edible but not to be recommended.

Range and habitat: A widespread species but generally rather uncommon. However, in suitable habitats it may be locally common over small areas. It is most frequently encountered among leaf-litter in Beech woods in southern England, especially where the soil is chalk.

Similar species: This species is difficult to confuse with any other because of its size and shape. *C. fistulosus* is extremely slender by comparison. It also occurs in Beech woods, growing on fallen twigs, and is rather uncommon.

GOLDEN SPINDLES

Clavulinopsis fusiformis

Height: 5-15cm

Characteristics: An eye-catching species which grows in dense tufts. The individual spindles are generally unbranched with a pointed tip. The spindles are usually rounded although they may be flattened and grooved towards the tip. The colour is a bright, golden yellow although the tip may turn brownish with age. Not edible. The flesh is whitish.

Range and habitat: A widespread and fairly common species throughout the region. It grows in grassy habitats, mainly on heathland, and often in large, dense tufts.

Similar species: *C. luteo-alba* has shorter (2-5cm), stouter spindles which are also unbranched. The colour is golden-yellow but the spindles have white tips. It is common and widespread in short grass. *C. helvola* has taller (3-6cm), slightly swollen spindles with a rounded tips and is found in small tufts in grassy places. A common and widespread species throughout the region.

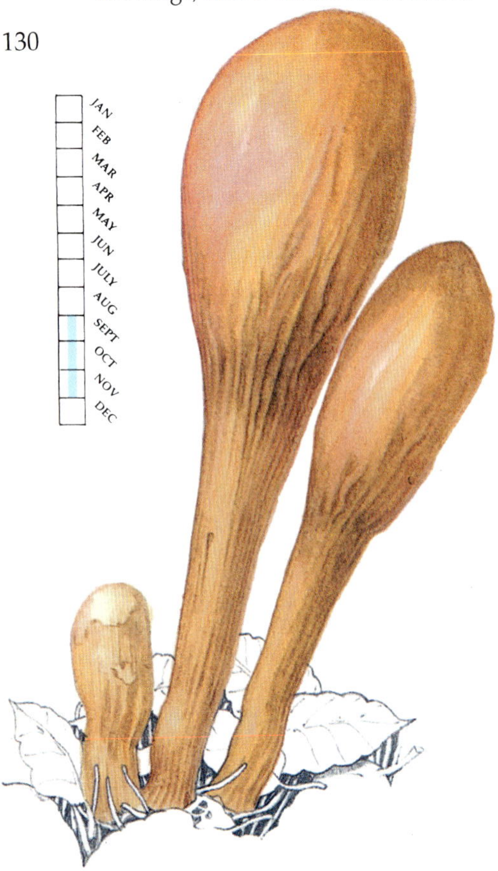

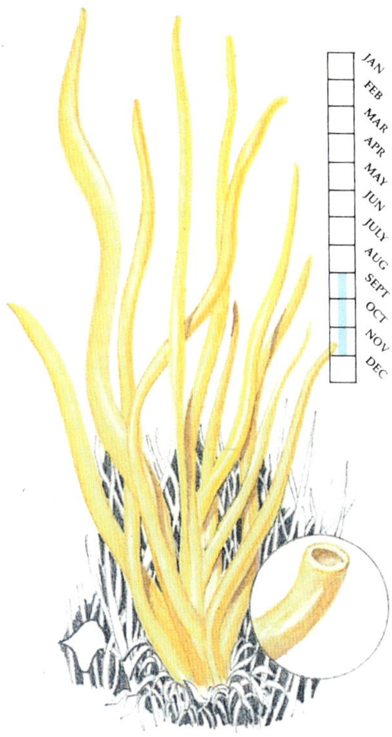

WHITE CORAL FUNGUS

Clavulina cristata
Height: 2-8cm
Characteristics: An attractive species which grows in tufts or small clumps. The individual clubs are branched and highly divided and the tips are fringed and pointed. The base of the stem is usually noticeably thickened. The colour is pure white although the tips may darken with age. Edible but not worth considering. The flesh is white.
Range and habitat: An extremely common and widespread species throughout Britain and northern Europe. It grows among leaf-litter and along forest rides and paths in both deciduous and coniferous woodland.
Similar species: The Grey Coral Fungus, *C. cinerea,* is very similar in appearance but ashy-grey in colour with blunter rather than pointed tips. The Wrinkled Club, *C. rugosa,* is white and solitary with the tips of the clubs blunted-ended and broad. As the English name suggests, the surface is wrinkled. Both species are common in woodlands.

CAULIFLOWER FUNGUS

Sparassis crispa
Width: 25-50cm
Characteristics: An extremely distinctive species which has a resemblance to a large, compact cauliflower head. The fungus comprises large numbers of densely packed, wrinkled lobes which arise from a central, rooting base. The surface texture is crisp and tough. The colour is greyish-buff to tan, darkening with age. Edible when young although it requires considerable preparation.
Range and habitat: A widespread species throughout the region which may be locally common in suitable habitats. It grows at the base of dying coniferous trees as well as beside rotting stumps.
Similar species: This species' size and unique appearance make it difficult to confuse with any other. *Ramaria botrytis* forms an almost globular head, up to 20cm across, which comprises closely packed, branched and pink clubs. It is a rather uncommon species found in deciduous woodland.

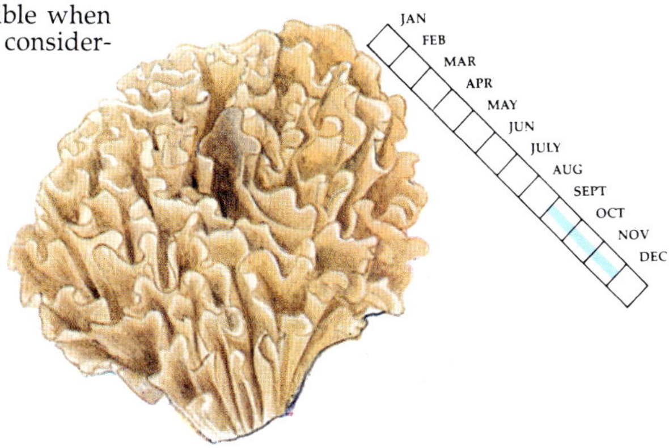

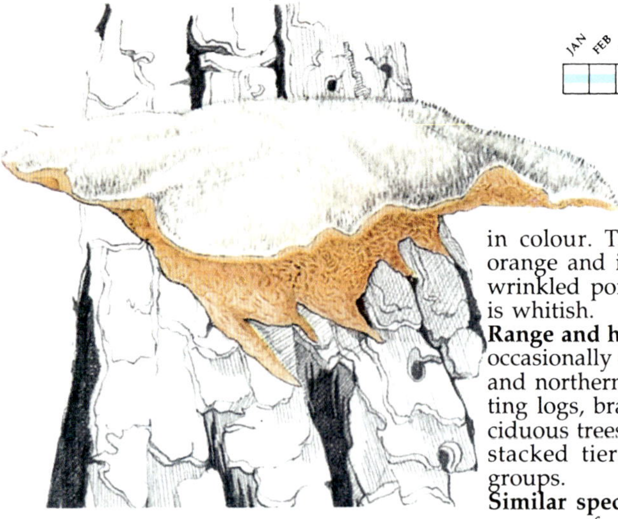

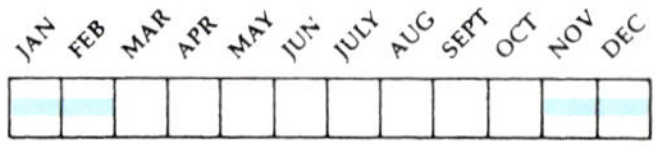

MERULIUS TREMELLOSUS

Bracket: 1-5cm across

Characteristics: The fruiting body forms soft, gelatinous brackets which are usually roughly semi-circular. They sometimes grow pressed close to the substrate. The upper surface is cloaked in woolly hairs and is whitish in colour. The underside is pinkish-orange and is covered in uneven and wrinkled pores. Not edible. The flesh is whitish.

Range and habitat: A widespread and occasionally common species in Britain and northern Europe. It grows on rotting logs, branches and stumps of deciduous trees and sometimes grows in stacked tiers or in large, spreading groups.

Similar species: *Coniophora puteana* is a cause of wet rot in housing but is widely found in the wild on decaying wood. The fruiting body forms extensive, olive-brown, warty patches on trunks and stumps. *Meruliopsis taxicola* forms red patches with a distinct white margin on coniferous timber. It is a rather uncommon species throughout the region.

PHLEBIA RADIATA

Diameter: 6-10cm

Characteristics: The fruiting body forms extensive, irregular to oval-shaped patches which often fuse together to cover much larger areas of bark. The surface is gelatinous, highly wrinkled and knobbly with radiating ridges and grooves. The colour is usually reddish-orange and is generally brightest around the margin. Some specimens are much duller. Not edible.

Range and habitat: A common and widespread species throughout Britain and northern Europe. It grows on the bark of deciduous trees and is especially frequent on Birch, Beech and Alder.

Similar species: Dry-rot Fungus, *Serpula lacrymans*, is common and widespread in wooded habitats. The fruiting body may form orange-yellow brackets although more usually it forms extensive patches. *Serpula himantioides* grows on coniferous timber and forms reddish-brown patches with a white margin. Widespread but not common.

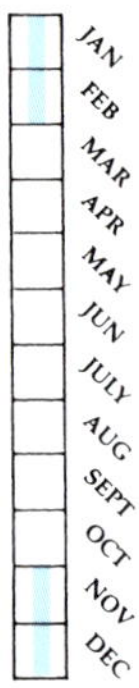

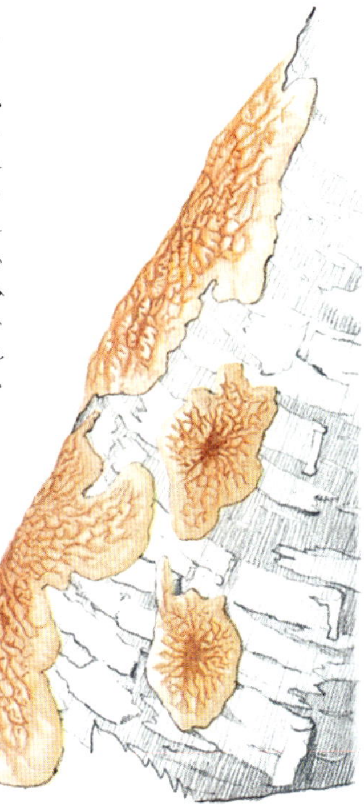

Calocera viscosa

JELLY FUNGI

The Jelly Fungi and their allies comprise a loose association of three related families, the Auriculariaceae, represented here by the Jew's Ear, *Auricularia auricula-judae*, the Tremellaceae, represented here by the Yellow Brain Fungus, *Tremella mesenterica*, and the Dacrymycetaceae, represented here by the Yellow Stagshorn Fungus, *Calocera viscosa*. They are all very different in appearance but share a gelatinous texture.

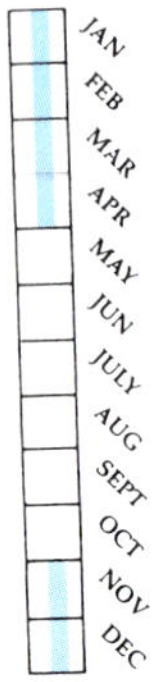

YELLOW BRAIN FUNGUS

Tremella mesenterica

Fruit body: 5-10cm across

Characteristics: A most distinctive and well-named species. The fruit body consists of an irregular gelatinous and slimy mass which is extremely folded and convoluted and resembles brain tissue. The colour is bright and eye-catching orange-yellow in fresh specimens. As the fungus dries and ages, it becomes dark orange-brown and is rather hard. Not edible.

Range and habitat: A common and widespread species in Britain and northern Europe which, because of its bright colour, is frequently encountered. It grows on dead twigs and branches of deciduous trees, both fallen and still attached to the tree itself.

Similar species: *T. foliacea* is brownish in colour and comprises more rounded and swollen lobes and folds forming a globular mass. It is common and widespread throughout the region and grows on dead twigs and branches of deciduous trees.

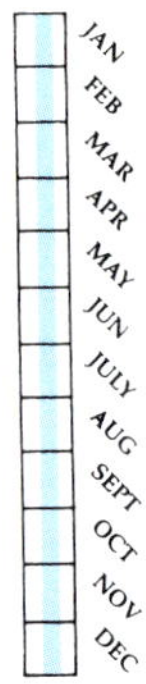

WITCHES' BUTTER

Exidia glandulosa

Fruit body: 2-6cm across

Characteristics: The fruit bodies of this species are supposed to resemble blackened knobs of butter. Individual fruit bodies comprise gelatinous blobs which may become convoluted and brain-like. The upper surface is dotted with black projections and the colour of the whole fruit body is a rich, dark brown. Not edible.

Range and habitat: A common and widespread species in Britain and northern Europe. It grows generally on dead twigs and branches of deciduous trees, especially on Oak. Several fruit bodies may fuse together in some instances to form large masses.

Similar species: The appearance and colour make it easy to identify. Compare with *Tremella foliacea*.

JEW'S EAR

Hirneola (Auricularia) auricula-judae

Fruit body: 4-8cm across

Characteristics: The fruit body is very variable in shape but is frequently ear-shaped. The English name derives from its supposed resemblance to Judas' ear who is reputed to have hanged himself on Elder, this species' usual host. The texture is gelatinous, the outer surface being slightly downy with the inner surface smooth. Fresh specimens are reddish-brown to tan, soft and sometimes partially translucent. As it dries and ages, the fungus becomes hard and dark. Edible and considered good by a few people.

Range and habitat: A common and widespread species throughout the region. It grows on branches of deciduous trees and especially on Elder.

Similar species: The Jelly Tongue, *Pseudohydnum gelatinosum*, grows on conifer stumps. The fruit body is tongue like, grey in fresh specimens but brown with age, with the underside covered in peg-like projections.

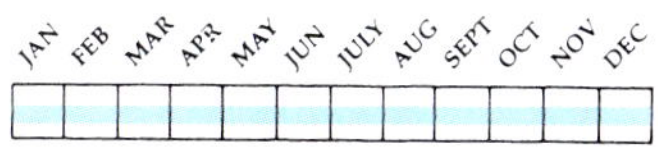

YELLOW STAGSHORN FUNGUS

Calocera viscosa

Height: 4-10cm

Characteristics: Although this fungus may appear superficially similar to several species of *Clavulinopsis*, its gelatinous and slimy texture should easily distinguish it. The fruit body comprises numerous tough, branched and antler-like projections which look like miniature stag's horns. They are slippery to touch and the colour in fresh specimens is bright golden-yellow or orange. In older specimens, the colour dries darker orange. Not edible.

Range and habitat: A common and widespread species in Britain and northern Europe. It grows on decaying conifer stumps and sometimes attached to partly buried roots. It is firmly rooted to the substrate.

Similar species: *C. cornea* has shorter (2-4cm high), unbranched clubs which are also gelatinous and orange-yellow in colour. It grows on stumps, twigs and branches of deciduous trees and is a common and widespread species.

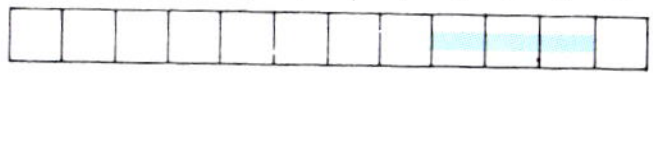

Geastrum triplex

PUFFBALLS AND THEIR ALLIES

This group of fungi comprises diverse and, in many cases, completely unrelated species. They have in common the fact that the spores are not produced on gills but rather on an outer surface coating. The range of shapes in this group is astonishing: from stinkhorns and puffballs to earth-stars and bird's-nest fungi. Each has its own special way of ensuring dispersal of the spores. Some use flies and other insects while others rely on the action of wind and rain. These are, indeed, some of the strangest of all our fungi.

STINKHORN

Phallus impudicus

Height: 15-25cm

Characteristics: One of our most distinctive and familiar fungi. The fruit body starts life as an egg-shaped capsule attached to buried, rotting wood by a tough mycelial strand. The capsule soon ruptures to reveal first a jelly-like layer and later the white, spongy stalk. The top is surmounted by a semi-globular head, honeycombed with pits and covered in a thick, greenish-black slimy mass which contains the spores. This mass has an extremely strong and repulsive smell which attracts the numerous flies that perform the dispersal of the spores. The smell can be detected by the human nose at a considerable distance. Within a matter of hours, most of the spore mass will have been removed by the insects.

Range and habitat: An extremely common and widespread species. It grows in wooded habitats in association with buried, rotting wood.

Similar species: *P. hadriani* is very similar but grows on sand dunes.

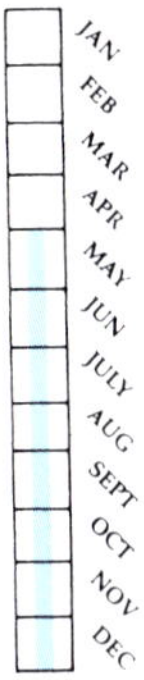

DOG STINKHORN

Mutinus caninus

Height: 10-12cm

Characteristics: Another curious and very distinctive species. The fruit body starts as an oval or cylindrical egg among the leaf-litter. The capsule soon ruptures and the slender, curved stalk emerges. This is buffish-white in colour, although some specimens are stained bright orange, and it has a pointed, conical head. This is bright orange in colour but covered in a dark greenish-black spore mass at first. This is sticky and slimy but has a smell which is mild in comparison to its common relative. The spore mass is removed in a matter of a few hours. Not edible.

Range and habitat: A common and widespread species in Britain and northern Europe. It grows among leaf-litter in deciduous woodlands, especially under Beech or Oak.

Similar species: The unique appearance and colour of this species make it difficult to confuse with any other. *Phallus impudicus* is much larger and stouter with a globular head.

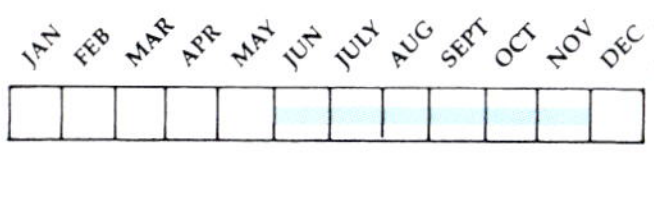

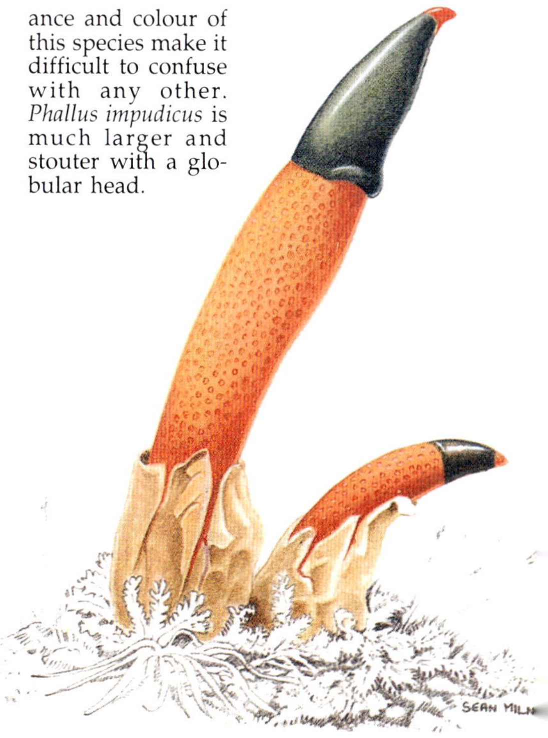

CAGE FUNGUS

Clathrus ruber

Diameter: 6-12cm

Characteristics: A most extraordinary species which starts as an off-white, egg-like capsule similar to that of the Stinkhorn, *Phallus impudicus*. The capsule soon ruptures to reveal an inflated, lattice-like spherical fruit body. This is bright red in colour on the inside and spongy and brittle to touch. The inside surface of the lattice is coated with a sticky slime which contains the spores. This is dark olive-brown in colour and has an unpleasant and far-carrying smell which attracts the dispersing agents – flies. Not edible.

Range and habitat: Occurs in gardens and woods. A rare and local species in Britain where it has been introduced. Naturalised in the Isles of Scilly and several spots in southern England but elsewhere only of casual occurrence.

Similar species: *C. archeri* has a bizarre appearance. From the egg-like capsule emerge 4-8 reddish 'arms', like a starfish. Naturalised in a few southern localities.

TULOSTOMA BRUMALE

Height: 2-3cm

Characteristics: A strange-looking species which comprises a globular head carried on a slender stem. The head is pale creamy-buff in colour and has central pore surrounded by a dark ring. The spores are released through this opening when the head is buffeted by the wind or pounded by raindrops. The stem is sometimes distorted and grooved and is covered in fibrous scales. Not edible.

Range and habitat: An uncommon and local species throughout Britain and northern Europe but one which is almost certainly overlooked. It has a distinctly southern distribution and is usually found growing on south-facing, sunny slopes especially where the underlying soil is sand or chalk. It has been found growing in mortar on old walls.

Similar species: *Battarraea phalloides* is much larger and comprises a brown spore-sac on a tall (20cm), shaggy stem. A rare and local species found in southern England on warm, well-drained soil.

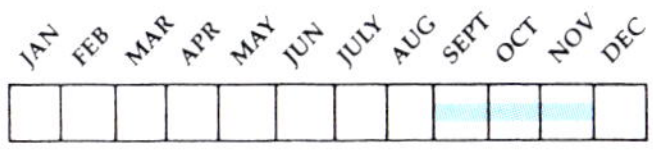

BIRD'S NEST FUNGUS

Cyathus striatus

Diameter: 0.5-1cm

Characteristics: The fruit body consists of a cone-shaped cap, the outer surface of which is covered in shaggy, fibrous scaly hairs. The inner surface is greyish-white and smooth and is usually radially grooved or lined. At the bottom of the cup lie several flattened 'eggs' which contain the spores. These are loosely attached to the cup by thin fibres. During downpours, raindrops cause these spore capsules to splash out and disperse. Not edible.

Range and habitat: A widespread species throughout the region but generally rather uncommon. It grows on stumps and fallen logs in woodland, often in large groups, and is very easy to overlook.

Similar species: *C. olla* has a grey-brown, slightly leathery outer surface to the trumpet-shaped cap. The inner surface and the 'eggs' are greyish. A locally common species, growing in open places and often on bare soil.

COMMON BIRD'S NEST FUNGUS

Crucibulum vulgare (=laeve)

Diameter: 0.5-1cm

Characteristics: An elegant and curious little species, the fruit body of which forms a neat cup or bowl. The outer surface of the cup is covered in fibrous, brown hairs and the inner surface is smooth and yellowish-grey. In the bottom of the cup lie 10 or more white and slightly flattened 'eggs' which contain the spores. They are attached to the cup by minute fibres. As with *Cyathus striatus*, the spore capsules are dispersed with the assistance of raindrops which hit the cup and bounce out with the 'egg'. Not edible.

Range and habitat: A widespread species in Britain and northern Europe but generally rather uncommon and local. It grows on logs, stumps and twigs, often occurring in large groups on suitable substrates. Probably frequently overlooked because of its small size.

Similar species: *Cyathus striatus* also grows on wood but has a radially grooved inner surface.

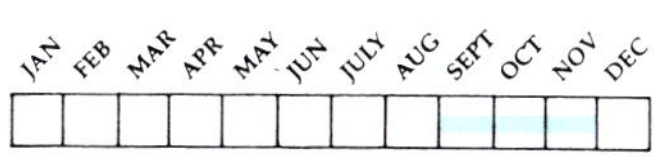

BOVISTA NIGRESCENS

Diameter: 4-8cm

Characteristics: The fruit body is a neat, round ball which lacks a basal stalk. In young specimens the surface is white but it turns brown with age. As the fungus matures, the outer layer of the fruit body usually peels away to reveal the black inner layer. In mature specimens, the ball splits and cracks irregularly at the apex and the spores are released, assisted by the action of wind and heavy rain. At first the fruit body is attached to the ground by a mycelial thread but this soon breaks allowing it to roll around, thus further assisting spore dispersal. Edible when young, before the spores mature.

Range and habitat: A widespread species throughout the region which is most common in the north and west of Britain. It grows on moorland and on upland grassland.

Similar species: *B. plumbea* is the southern and lowland equivalent of *B. nigrescens* and very similar in appearance. It grows in meadows, lawns, parks and on golf courses.

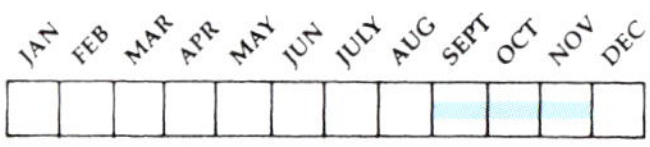

COMMON PUFFBALL

Lycoperdon perlatum

Diameter: 3-6cm

Characteristics: The fruit body comprises a globular head on a short, stout stem which is firmly attached to the substrate. The head in particular is covered in warty protuberances which wear off with age leaving a knobbly, patterned surface. The colour is creamy-white, sometimes rather buffish on the globular head. In mature specimens, a pore opens in the apex of the head through which the spores escape. When the puffball is battered by rain or knocked, clouds of spores are expelled and carried by the wind. Edible when young.

Range and habitat: A common and widespread species throughout the region. It grows among leaf-litter in woodlands and often occurs in groups.

Similar species: *L. echinatum* is similar in size and shape and also grows in woodland. However, the colour is rich-brown and the surface is covered in spiny scales which gradually become detached and worn from the apex downwards.

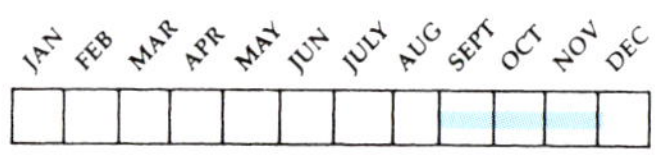

Bovista nigrescens

Lycoperdon perlatum

PESTLE PUFFBALL

Calvatia excipuliformis

Diameter: 4-10cm

Height: up to 15cm

Characteristics: The fruit body is a rounded, globular head on a long and stout stem. The English name derives from its resemblance to a pestle. The surface of the head is covered in granular, warty flecks which soon rub off leaving a fibrous texture. In young specimens, the colour is creamy-white but this becomes brownish with age. The skin peels and splits at the apex with age to liberate the spores. These are expelled when the puffball is battered by wind and rain. The empty, tattered remains persist for a long time. Edible when young.

Range and habitat: Common and widespread. It grows along woodland rides and on heaths.

Similar species: *Calvatia utriformis* is a more compact species with a short, broad stem. The surface is creamy-white at first, turning brown with age. It is covered in warts at first which soon rub off. Grows in fields and meadows.

GIANT PUFFBALL

Langermannia gigantea

Diameter: 10-70cm

Characteristics: When fully formed, this is one our most unmistakable species as well as our largest. The fruit body is often the size of a football. It is rounded but often slightly distorted and creased, especially towards the base. Before maturity, the fungus is creamy-white in colour and leathery. As the spores mature, the outer skin peels away to reveal the spore mass. The whole fruit body turns brown with age and the mycelial 'root' often breaks so that the puffball can roll around freely. Edible and delicious when young. Completely inedible once the spores have begun to mature.

Range and habitat: A widespread species but rather local. In suitable habitats, however, it may be abundant. It grows in meadows and pastures and beside grassy, overgrown hedgerows.

Similar species: The enormous size of mature specimens of this species make it almost impossible to confuse with any other.

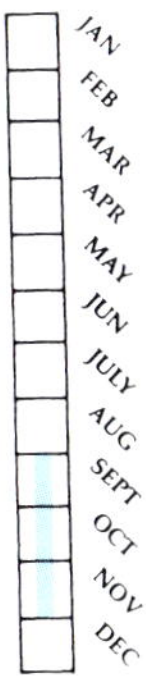

Calvatia excipuliformis

Langermannia gigantea

STUMP PUFFBALL

Lycoperdon pyriforme

Diameter: 2-4cm

Height: 3-5cm

Characteristics: The fruit body is rather club-shaped or pear-shaped comprising a rounded head carried on a distinct stem. The colour is creamy-white but fades to brown with age. The surface is covered by a layer of warty scales, most dense on the head, which soon rub off to leave a smooth skin. The fruit body is attached to the substrate by a mycelial strand. As the puffball matures, an apical pore opens through which the spores are expelled by the action of wind and rain and when the fruit body is knocked. Edible when young. Spores olive-brown.

Range and habitat: A common and widespread species. It grows on rotting stumps and logs and half buried, decaying wood. This is the only puffball likely to be found on this habitat. Grows in large groups.

Similar species: *Vascellum pratense* is common in grassy places. It is 2-4cm across, creamy-white turning brown, and the surface has only small scales.

COMMON EARTHBALL

Scleroderma citrinum

Diameter: 4-10cm

Characteristics: The fruit body is an irregularly rounded sphere, sometimes shaped rather like an apple. It is attached to the soil by a tough mycelial strand. The outer surface is olive-brown to buff and is covered with rough, flaky scales. As the fruit body matures, it splits irregularly and spores are expelled through the gaping hole. Their liberation and dispersal is assisted by the action of wind and rain. Not edible and may even be poisonous.

Range and habitat: An extremely common and widespread species in Britain and northern Europe. It grows in a variety of wooded habitats from leaf-litter to paths and rides and also on scrubby heathland. In suitable areas, the woodland floor may seemingly be covered with earthballs.

Similar species: *S. areolatum* is similar but 1-4cm across. It has a smoother skin with finer scales and grows along grassy, woodland rides. It is widespread but less common.

EARTHSTAR

Geastrum triplex

Diameter: 4-10cm

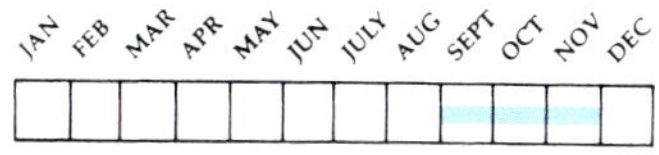

Characteristics: A most extraordinary species. At first, the fruit body looks like a hyacinth bulb in the leaf litter. The surface is tough and fibrous but the outer layer soon splits into 4-8 radiating arms which fold back and resemble a star. The inner surface of these arms is smooth and grey-brown and may crack as the tissue bends. As the arms fold back they reveal the inner ball-like sac which contains the spores. These are expelled through the apical pore when the fungus is buffeted by wind and rain or when knocked. The spore sac sits in a collar of leathery tissue. The radiating arms may fold back sufficiently to lift the spore sac clear of the ground. Not edible.

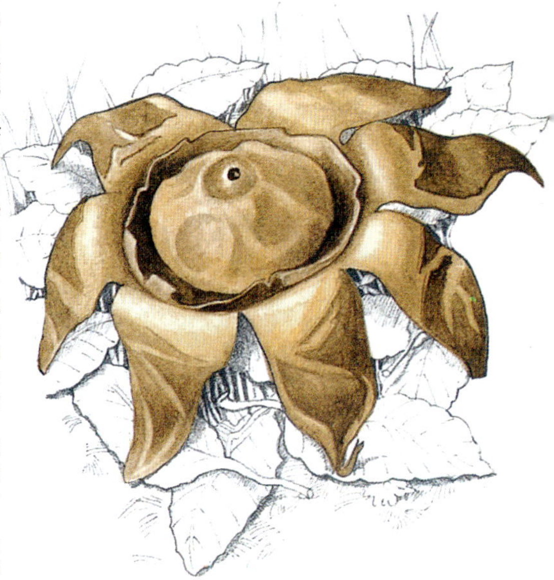

Range and habitat: A widespread but local species throughout the region. It grows in leaf litter in deciduous woodland.

Similar species: *G. rufescens* is similar but lacks the collar beneath the spore sac. Grows in leaf-litter.

SCLERODERMA VERRUCOSUM

Diameter: 3-6cm

Characteristics: The fruit body is spherical to egg-shaped and is attached to a long, tough and tapering stem which is strongly rooting. The outer surface of the fruit body is pale olive-brown and is covered in wart-like scales. The stem is paler and smoother and usually ribbed and twisted with soil attached. When the fungus matures, the apex to the fruit body splits and cracks irregularly. The spores are expelled when battered by wind and rain or knocked. Not edible.

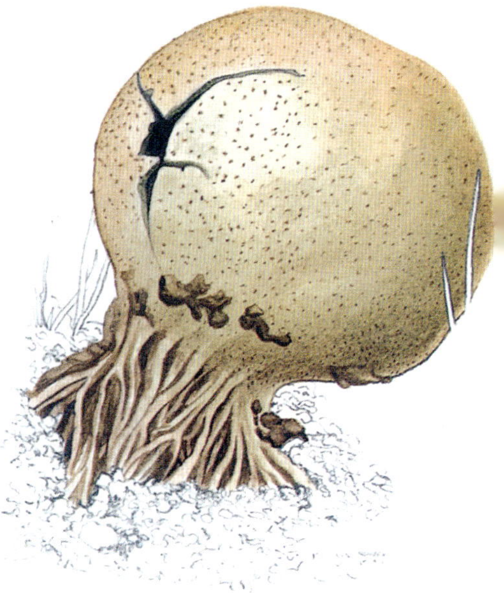

Range and habitat: A common and widespread species throughout Britain and northern Europe. It grows in wooded habitats and heathlands, usually on sandy soils.

Similar species: *Lycoperdon foetidum* is similar in appearance. The surface is brown and covered in warty scales which rub off. The rooting stem is not elongated and the spores are liberated through an apical pore. A common and widespread species which grows on heaths and in woodlands.

Helvella crispa

CUP FUNGI AND THEIR ALLIES

Among the larger fungi, the cup fungi, or *Discomycetes*, comprise the major part of a group known as the *Ascomycetes*, the remaining members of which are the flask fungi, or *Pyrenomycetes*. Although some of the cup fungi, and especially the morels, may superficially resemble more 'conventional' mushrooms and toadstools (*Basidiomycetes*), this apparent similarity is quite misleading. The difference lies in their method of spore production and the way in which these are liberated. In addition to the morels, the cup fungi members include *Helvellas*, Elf Cups, Earth Tongues and Batchelor's Buttons: the intriguing English names reflecting their strange forms.

MOREL

Morchella esculenta

Height: 5-20cm

Characteristics: A well-known and extraordinary species. The head is variable in shape from rounded to irregularly oval. The surface comprises a honeycomb of sharp ridges and deep pits and is olive-buff to brown in colour. The texture is sponge-like and the head is generally hollow. The stem is stout and often hollow. It is white and often swollen towards the base. Edible and delicious. This a highly prized species.

Range and habitat: A widespread species throughout the region but seldom common and rather unpredictable in appearance. It grows generally on chalky soil in grassy woodlands, field margins, hedgerows and roadside verges.

Similar species: *M. vulgaris* has a more rounded, brownish-buff cap and grows in similar habitats. The convoluted ridges have rounded edges. *M. conica* has distinctly conical head which is rich brown in colour.

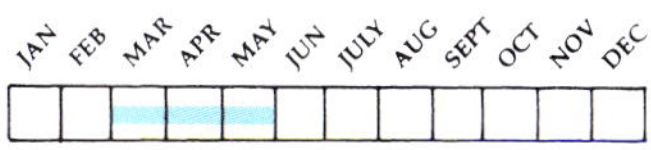

FALSE MOREL

Gyromitra esculenta

Height: 6-12cm

Characteristics: The cap or head of this strange-looking species is a large, irregular mass which comprises convoluted lobes and folds. These may resemble brain tissue in mature specimens. The colour is reddish-tan to orange-brown. The stem is short, stout and often distorted. The stem colour is whitish. Cutting the stem reveals numerous hollow chambers. Extremely poisonous although apparently edible after boiling. Best avoided.

Range and habitat: A widespread species in Britain and northern Europe but everywhere rather uncommon. It grows on well-drained, sandy soils, sometimes among pine needles in coniferous woodland but it occasionally occurs on sand dunes.

Similar species: *G. infula* is an uncommon woodland species with a smaller, less convoluted brown head and a more slender, white stem. Compare both species with the true Morels.

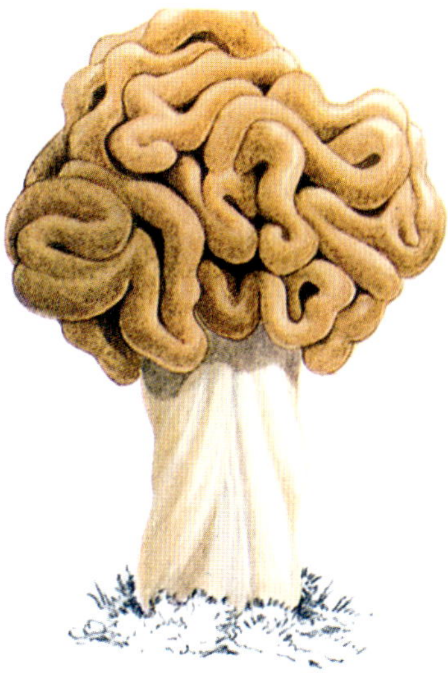

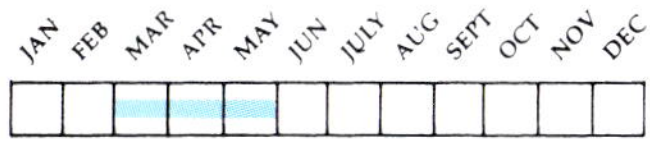

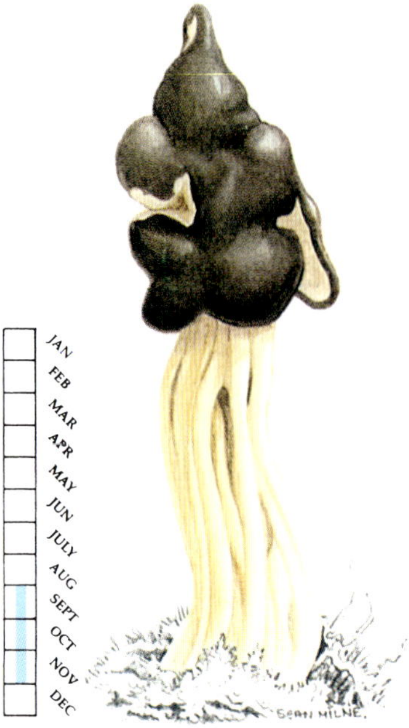

BLACK HELVELLA

Helvella lacunosa

Height: 4-8cm

Characteristics: The cap is an irregular saddle-shape which is generally unevenly distorted and folded back over the stem. One lobe of the cap may be partially twisted and point skywards. The upper surface of the cap is blackish-grey while the underside is usually slightly paler. The stem is whitish, sometimes much darker, with longitudinal grooves and folds and may appear to comprise fused strands or fibres. Edible but generally not worth considering.

Range and habitat: A widespread species in Britain and northern Europe which is locally common in some areas. It grows in both coniferous and deciduous woodland and appears to favour areas of burnt soil.

Similar species: *Leptopodia atra* is smaller (4-6cm tall) with a less distorted, greyish-brown cap and a more slender, brown stem. A rather uncommon species which is found in both coniferous and deciduous woodland.

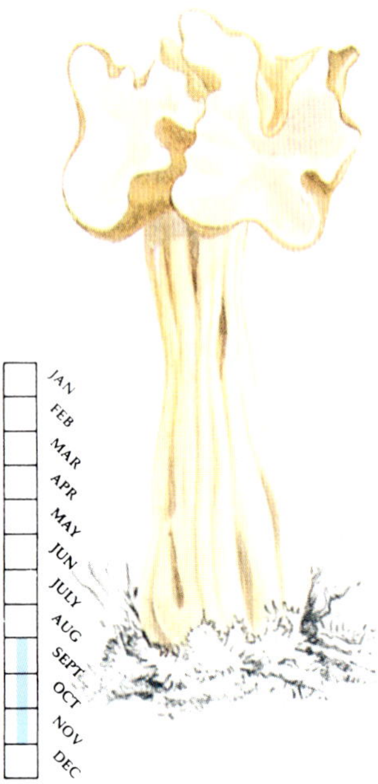

WHITE HELVELLA

Helvella crispa

Height: 6-10cm

Characteristics: A distinctive and unusual species. The cap is flattened, folded and saddle-shaped, often with one of the folds pointing skywards. The surface is often convoluted towards the apex. The colour of the upper surface is creamy-white while the underside is greyish-buff. The pure white, hollow stem is extremely furrowed and fluted along its length and often rather distorted. Edible but not worth considering.

Range and habitat: A widespread species in Britain and northern Europe which is locally common in some areas. It grows in a variety of deciduous woodland habitats from grassy rides and verges to forest rides.

Similar species: *Leptopodia elastica* is up to 8cm tall and has buffish-brown, saddle-shaped cap with a long, slender stem which is white in colour. A rather uncommon species throughout the region which occurs in wooded habitats. Edible but not worth considering.

ORANGE PEEL FUNGUS

Aleuria aurantia

Diameter: 1-8cm

Characteristics: The English name gives a perfect description of this species. Indeed, it may easily be passed over as discarded orange peel. It forms irregularly shaped and sometimes partially flattened cups, often several 'scattered' over a small area. The inner surface is a bright golden-yellow or orange and smooth. The outer surface is whitish and has a powdery texture. Not edible.

Range and habitat: A common and widespread species throughout Britain and northern Europe. It grows in a variety of habitats from deciduous woodland floors to lawns and roadside verges. Often grows on bare soil along paths and tracks, a habitat which may reinforce the impression that the fungus is just discarded peel.

Similar species: The combination of bright orange inner surface and whitish outer surface make this species difficult to confuse with any other. However, compare with some species of *Peziza*.

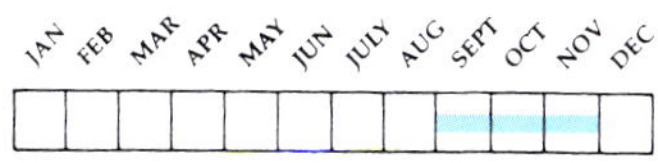

PEZIZA REPANDA

Diameter: 4-10cm

Characteristics: The fungus is an irregular cup-shape at first. With age, it flattens and the margin may tear and fray, becoming rather serrated. The inner surface is orange-brown to buff and the outer surface is whitish and slightly powdery. The cup may be attached to the substrate by a short stem. Not edible.

Range and habitat: A widespread species in Britain and northern Europe which is locally common in suitable habitats. It grows in mature gardens and deciduous woodlands, generally around stumps or near buried wood.

Similar species: *P. badia* forms irregular cups, 4-8cm across, with undulating margins. The inner surface is dull brown and the outer surface is reddish brown. A common and widespread species which grows in bare and open situations on paths, tracks in hedgerows. *P. echinospora* is similar but has the outer surface extremely powdery. A common and widespread species which grows in open situations, often on burnt soil.

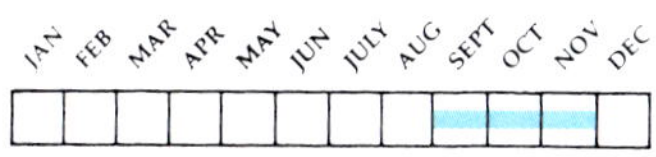

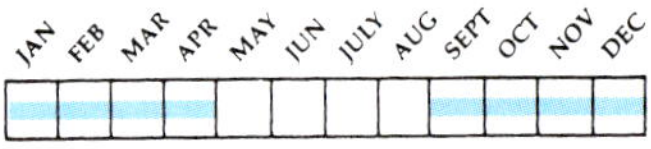

PEZIZA VESICULOSA

Diameter: 4-8cm

Characteristics: This cup-shaped species has a smoothly wrinkled surface. The margin is distinctly inrolled and is usually frayed or toothed. The cup seldom flattens or expands and so retains its rather spherical shape with age. The inner surface is orange-buff, smooth then blistering with age, while the outer surface is pale buff and powdery to touch. Not edible and may even be poisonous.

Range and habitat: A common and widespread species in Britain and northern Europe. It grows on manure, compost heaps, and rotting bales of straw.

Similar species: The Pine Fire Fungus, *Rhizina undulata*, forms rich-brown, lobed cushions on coniferous woodland floors, especially in areas where the soil has been burnt. It is rather uncommon. *Paxina acetabulum* is an uncommon woodland species which grows on chalky soil. The chestnut-brown cup is supported on a whitish-buff stem with branched ribs.

SCARLET ELF CUP

Sarcoscypha coccinea

Diameter: 2-5cm

Characteristics: This species forms beautiful little cups or bowls, attached to the substrate by short stems. The inner surface is bright red or scarlet and smooth while the outer surface is whitish and downy or slightly hairy. With age, the margin of the cup becomes rather tattered and frayed. Edible but not worth considering.

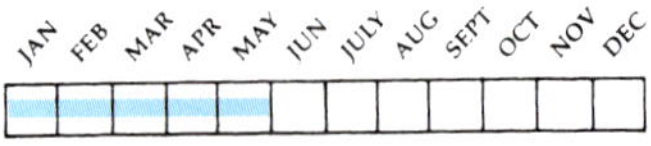

Range and habitat: A widespread species in Britain and northern Europe. Rather local and perhaps declining. Most frequently found in the west of Britain. It grows on branches and twigs on the woodland floor. Several cups may be found together on suitable substrates.

Similar species: The Eyelash Fungus, *Scutellinia scutellata*, forms red cups, 0.5cm in diameter, on dead wood in damp situations. The cup margin is fringed with black hairs and in the under surface is brown. A common and widespread species throughout the region.

Pigs looking for truffles

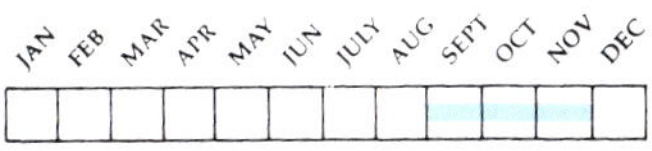

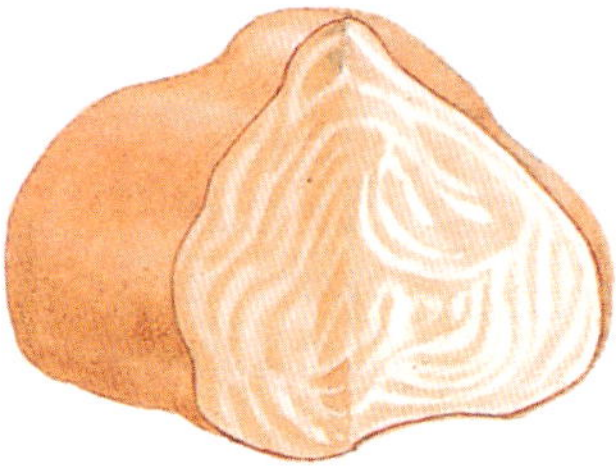

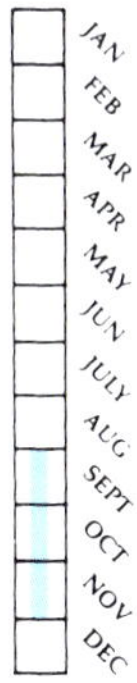

RED TRUFFLE

Tuber rufum

Diameter: 0.5-2cm

Characteristics: Like other truffles, the Red Truffle lies buried in the soil and is consequently difficult to find. It forms an irregular, reddish-orange mass which has a slightly warty surface. When cut, the inside flesh is greyish-white at first but gradually stains pinkish. It is lined with darker veins. Edible and good. This species has a nutty taste when fresh but with age acquires a rather unpleasant smell.

Range and habitat: A widespread species throughout the region but rather scarce. This is due in part to it being difficult to locate. It grows in Oak woodland, buried in the soil.

Similar species: The Truffle-like *Choiromyces meandriformis* looks rather like a knobbly Potato or a Jerusalem Artichoke. The skin is yellowish-brown in colour and the flesh is whitish. It lies at the surface of woodland soils and is widespread although rather local and uncommon. An edible species but not worth considering.

WHITE TRUFFLE

Tuber aestivum

Diameter: 3-6cm

Characteristics: This species is often almost spherical in shape, its surface covered in blackish-grey, angular warts. When cut, the flesh is white at first but gradually becomes marbled with fine, dark lines. Edible and delicious but not the most prized species.

Range and habitat: A widespread species in the region which, in Britain, is most usually encountered in the south. It is nowhere common but is probably much overlooked. It grows buried in the soil, generally in Beech woods on chalky soil. Its dark outer skin only adds to the difficulty in finding it.

Similar species: The Perigord Truffle, *T. melanosporum*, does not occur in Britain. It grows in Oak woods and is superficially similar to *T. aestivum.* The outer layer is covered in irregular, dark-brown to black warts and the flesh is marbled dark brown. The taste is outstanding and the truffles fetch a high price.

HARE'S EAR

Otidea leporina

Height: 4-10cm

Characteristics: A most distinctive species which lives up to its English name and resembles a hare's ear in size and shape, if not in colour. The fungus forms an off-centre and lop-sided, irregularly-shaped cup, the margin of which may become frayed and tattered with age. The inner surface is pinkish orange and smooth while the outer surface is bright yellow and rather powdery. A short stem attaches the fungus to the substrate. Not edible.

Range and habitat: A widespread species throughout the region but rather local and uncommon. It grows in leaf-litter in deciduous woodlands.

Similar species: *O. alutacea* is more cup-shaped although still lop-sided with a wavy margin. The inner surface is buffish-orange in colour while the outer surface is pale buff. It grows in bare soil in deciduous woodlands. A widespread species throughout the region but rather local and uncommon. Not edible.

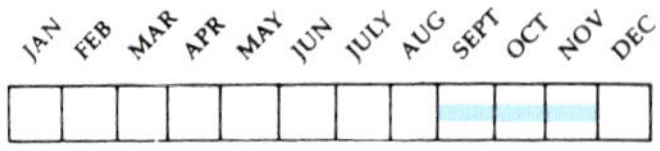

ASCOCORYNE SARCOIDES

Diameter: 0.5-1cm

Characteristics: At first this fungus forms a fleshy, gelatinous mass with a convoluted and irregular surface. Mature fruit bodies form in time and comprise button-like discs raised on short stalks which are sometimes rather cup-shaped. The margins are generally rather unevenly lobed. The colour of the fruit body is reddish-purple. Not edible.

Range and habitat: A common and widespread species in Britain and northern Europe. It grows on the decaying stumps, logs and branches of deciduous trees. It will sometimes form extensive masses on suitable substrates.

Similar species: *Neobulgaria pura* is rather similar and forms closely-packed masses of gelatinous fruit bodies, each of which is flat-topped or slightly concave. The colour is reddish-orange. A widespread species throughout the region which is rather uncommon. It grows on dead twigs and logs of deciduous trees, especially on Beech. Not edible.

BATCHELOR'S BUTTONS

Bulgaria inquinans

Fruit body diameter: 1-4cm

Characteristics: The fruit body forms a rubbery blob which is blackish brown in colour. In young specimens the margin may be slightly inrolled. With maturity, the fruit body expands and flattens with the outer surface being rather scurfy and the flattened disc is smooth and shiny. The fruit bodies often grow in closely-packed groups presenting an almost continuous, shiny exterior face rather like a row of buttons. Not edible. The spores are dark brown.

Range and habitat: A common and widespread species in Britain and northern Europe. It is found in deciduous woodland growing on dead twigs and branches. Most frequently associated with Oak and it may be found in large masses on suitable substrates.

Similar species: Rather similar to Witches' Butter, *Exidia glandulosa*, which has the disc dotted with warty scales rather than appearing as a smooth, shiny surface.

GREEN WOOD-CUP

Chlorociboria aeruginascens

Cup diameter: 1-4mm

Characteristics: The fruit bodies comprise small cup-like structures which are blue-green in colour. The cup is often irregularly shaped with a wavy margin and the centre is flattened or slightly depressed. The cups are supported on short stalks. The fruit bodies are less familiar than the effect which the mycelium has on its wood substrate. It stains the twigs or branches bright green. Not edible.

Range and habitat: A common and widespread species throughout the region. It grows on fallen wood of deciduous trees, especially on Oak, and green-stained twigs and branches are a familiar sight on many woodland floors.

Similar species: The yellow fruit bodies of *Bisporella citrina* are disc-shaped and grow in large masses on fallen branches and twigs of deciduous trees. A common and widespread species throughout the region. The mycelium does not stain wood.

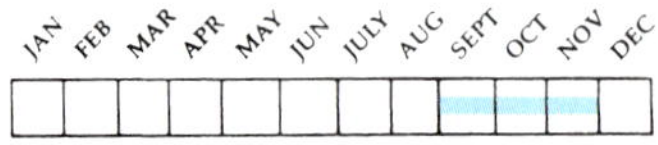

EARTH-TONGUE

Trichoglossum hirsutum

Height: 4-8cm

Characteristics: A curious-looking species which may be easily overlooked because of its appearance and growing habitat. The fruit body comprises an oval, tongue-shaped head on a long, slender stem. The head is rather flattened and grooved and may have a pointed tip in some specimens. The fruit body is black in colour and the stem, in particular, is covered in stiff, black hairs giving a velvety appearance. Not edible.

Range and habitat: A widespread species which is locally common in suitable habitats. It grows in grassy and mossy situations, in lawns and fields, and is very easy to overlook.

Similar species: *Geoglossum cookeianum* is superficially similar but more slender and tapering toward the narrow stem. It is 3-6cm high and the stem lacks the velvety coating of black hairs found in *Trichoglossum hirsutum*. It is a widespread species in the region but rather local, growing in short grass on sandy soil.

GREEN EARTH-TONGUE

Microglossum viride

Height: 2-5cm

Characteristics: As the English name suggests, the fruit body forms bright-green to olive-coloured tongues. The oval- or club-shaped head is often rather flattened and grooved and smooth. The stem is slender and bright-green, fading towards the base. It may have a granular surface. Not edible.

Range and habitat: A widespread species in Britain and northern Europe but only locally common. It grows among leaf litter and mosses in deciduous woodland.

Similar species: *Mitrula paludosa* is a curious-looking species with a bright orange-yellow, smooth head on a long, slender stem which is white. It grows up to 4cm high and is found in moist, mossy places in ditches or damp woodland. *Leotia lubrica* has a gelatinous, slimy head which is olive-green and rounded. It is carried on a buff stem. It grows in damp woodland and is rather local and uncommon throughout the region.

FLASK FUNGI

The flask fungi, or *Pyrenomycetes*, are Ascomycete fungi. Most members of this group are small and pass unnoticed but the few included in this book are comparatively large and conspicuous. Most are rather tough and woody although extremely varied in form. They range from the slender Candle Snuff Fungus, *Xylaria hypoxylon*, to the compact and round King Alfred's Cakes, *Daldinia concentrica*.

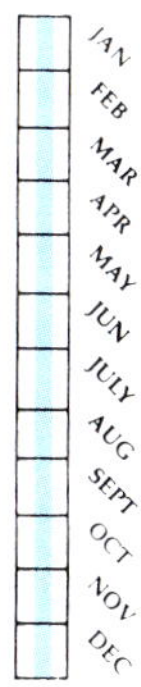

CANDLE-SNUFF FUNGUS

Xylaria hypoxylon

Height: 1-6cm

Characteristics: An aptly named species which, when mature, strongly resembles the snuffed-out wick of a candle. At first, the fruit body is a simple spike but with maturity it expands, flattens and branches and eventually looks like antlers. The lower half of the fruit body is blackish and downy. The branched portion is white and powdery but this is lost with age. Not edible.

Range and habitat: An extremely common and widespread species throughout the region. It grows on dead and decaying stumps, logs and branches in deciduous woodland. Often forms large masses on suitable substrates.

Similar species: The colour and shape of this species make it difficult to confuse. The Caterpillar Fungus, *Cordyceps militaris*, is an unusual species. It forms orange, clubbed spikes which grow from the pupae and pupating caterpillars of moths buried in the soil.

DEAD MAN'S FINGERS

Xylosphaera polymorpha

Height: 4-8cm

Characteristics: The fruit bodies form clumps of club-shaped spikes which are often rather rounded and finger-like. The upper portion is swollen and often irregular in shape. It has a blackish-brown surface which is rather rough and wrinkled. The flesh is white. The stem-like lower portion of the spike is more slender and paler. Not edible.

Range and habitat: A widespread species throughout Britain and northern Europe which can be common in suitable habitats. It grows in deciduous woodland, especially under Beech, and usually on rotting stumps. It may form large clumps on suitable substrates.

Similar species: *Cordyceps ophioglossoides* is a strange species which parasitises the truffle-like, subterranean fungus *Elaphomyces muricatus*. A club-shaped fruit body emerges from the host, the base of which is yellow while the swollen upper portion is black. Uncommon but overlooked on the woodland floor.

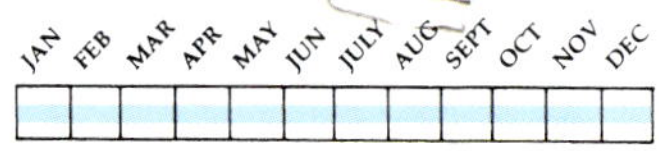

KING ALFRED'S CAKES; CRAMP BALLS

Daldinia concentrica

Diameter: 3-10cm

Characteristics: A distinctive species, the fruit body of which is irregularly rounded and sometimes semicircular in shape. The texture is extremely hard and woody and the colour is brownish-black at first but shiny black with age. When sectioned, the flesh appears marked with concentric zones of buff and dark brown. The English names derive from the burnt appearance of the fruit body and the belief that it was a cure for cramp. Old fruit bodies persist for a considerable time. Not edible.

Range and habitat: A common and widespread species in Britain and northern Europe. It grows on dead branches and stumps of deciduous trees and is most frequently associated with Ash and Beech. On suitable substrates, many fruit bodies may grow closely packed together.

Similar species: The distinctive appearance and hard texture of this species make it difficult to confuse with any other.

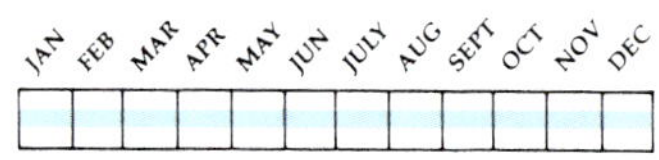

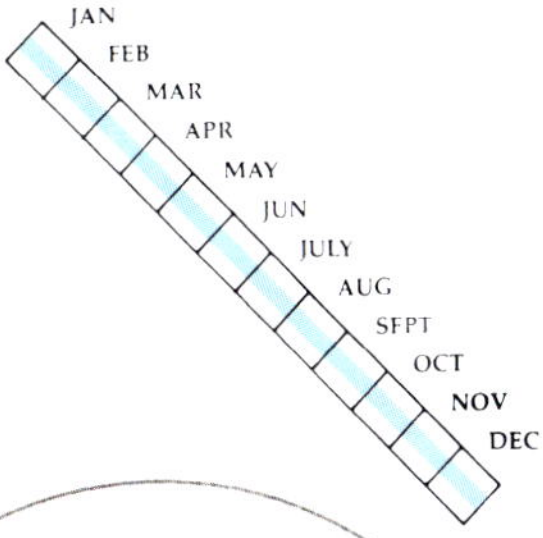

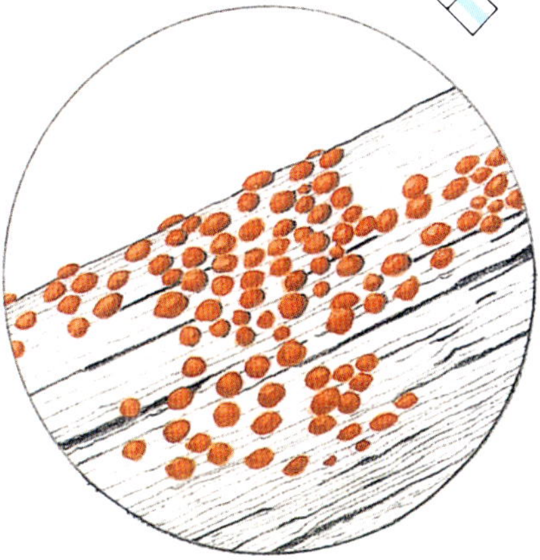

CORAL SPOT FUNGUS

Nectria cinnabarina

Fruit body: 1-3mm

Characteristics: The tiny fruit bodies invariably occur in large patches, dotted over the surface of bark and wood. Two forms of this species can be seen, often close together. The asexual stage forms soft, pinkish-orange cushions. The sexual stage forms cinnabar-red perithecia which are hard. The fruit bodies often push through the bark. Not edible.

Range and habitat: An extremely common and widespread species in Britain and northern Europe. It forms patches of orange spots on the bark of fallen twigs and branches, stumps and logs. It occasionally occurs on living wood.

Similar species: There are several other species of *Nectria* in the region. *N. peziza* is a yellow species which occurs on rotting wood and sometimes on old bracket fungi. *Diatrype disciformis* forms blackish disc-like spots, with white flesh, which grow in large patches on the bark of deciduous trees, and especially on Beech.

HYPOXYLON FRAGIFORME

Fruit body diameter: 0.5-1.5cm

Characteristics: The fruit body is hemispherical and domed, growing on the surface of bark. They may be rather irregular in shape, especially if growing in close proximity to other fruit bodies. The surface is rather rough and the colour is first orange-red then darkening to brick-red and finally blackening. Not edible.

Range and habitat: A common and widespread species in Britain and northern Europe. It grows on the surface of fallen and dead wood and is especially frequent on Beech. On suitable substrates, large patches may form.

Similar species: *H. nummularium* forms large, blackish patches rather than rounded fruit bodies. It is widespread but rather uncommon. It grows on dead wood of deciduous trees and especially on Beech. *Ustulina deusta* forms irregular, roughly-circular cushions which encrust decaying tree stumps. Concentric colour zones may form giving the fungus a superficial resemblance to a lichen.

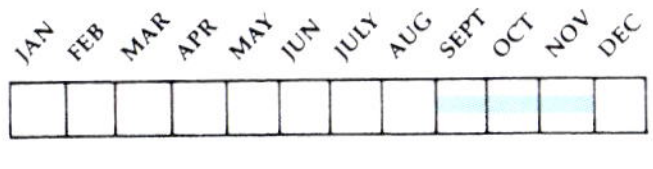

GLOSSARY

adnate gills or pores attached to the stem for most of their width
adpressed scales pressed or stuck onto the surface of the cap
apothecium the cup-shaped fruit body of a Discomycete fungus
ascus a cell in which the spores of Ascomycete fungi are produced
basidium a club-shaped cell on which the spores or Basidiomycete fungi are formed
bulbous distinctly swollen
campanulate bell-shaped
cap usually a flattened part of the fruit body of a mushroom or toadstool borne on a stem and under which the gills or tubes are suspended
convex a cap which is rounded or curved on the outside
cuticle the skin-like outer layer of some caps and stems
decurrent gills or tubes which run down the stem
deliquescent flesh, especially cap and gills, which liquifies with maturity
dentate toothed

Longitudinal sections of caps to show terms used for their shape and gill attachment.

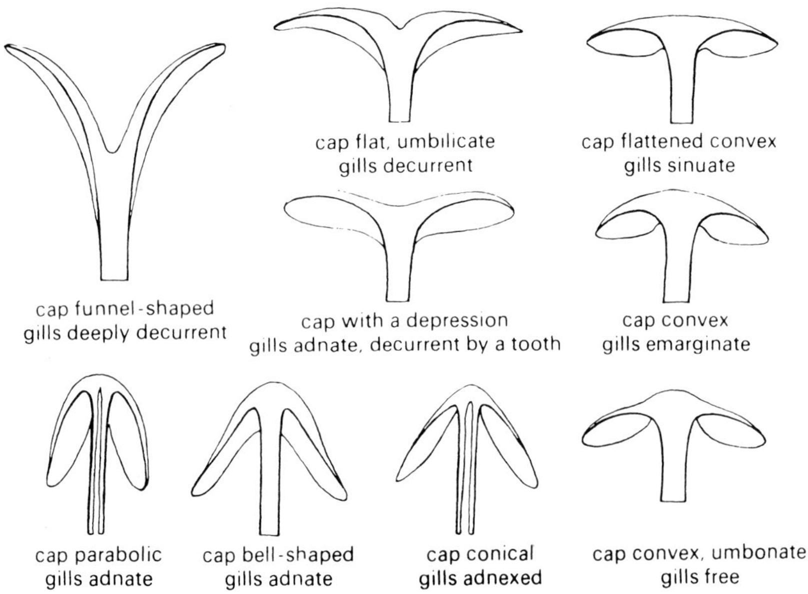

depressed sunken centre to the cap
distant gills widely spaced
fibrillose covered in fibres
fruit body the entire reproductive body or a fungus, for example the whole toadstool
gill plate-structure on the underside of the cap which produces spores
glutinous covered in a sticky or slimy coating
hyphae microscopic threads which comprise the mycelium
incurved curving inwards
inrolled curving downwards or inwards and rolling
mealy with a covering of powdery granules
mycelium a mass of hyphal threads which form a web-like mass
mycorrhiza a symbiotic relationship between the roots of a plant and the mycelium of a fungus
ochre pale yellow-brown
pilose covered in shaggy hairs
pruinose with a dusty coating
recurved curved back on itself
ring membranous tissue around the stem of some species of fungi
scaly covered in flakes or scales
spore microscopic reproductive body produced by the fruit body
stipe the stem of the fruit body
striate marked with grooves or line
tubes spore-producing structures found on the underside of the cap in species of Boletus and others
umbo a raised area in the centre of the cap
umbonate possessing an umbo
veil the membrane which protects the young fruiting body or its gills
viscid slimy and wet
volva sac-like structure which encloses the stem base in some species
waxy texture and appearance of candle wax

Acknowledgements

All photographs supplied by Nature Photographers unless otherwise stated. F Blackburn 16; R Bush 42,111; B Burbidge 11; A Cleave 37, 56, 149; C H Gomershall 123; D Hawes 28; M E Hems 15; J Hyett 51, 133; E a Janes 9, 17, 76, 82, 102; L G Jessup 129; D Osborn 20, 116; P Sterry 8, 13, 18, 19, 22, 38, 45, 74, 105, 124, 134, 136, 144, 153. Gordon Dickson/Natural Image 33.

Line drawings by Laura Mason.
Colour artwork: Roger Gorringe/Hamlyn Publishing Group 21top, 24top, 29top, 31bottom, 34bottom, 96top, 104bottom, 106top and bottom, 110bottom, 118top and bottom, 119top. All other artworks by Sean Milne/Equinox Limited.

INDEX